MIMO Communication for Cellular Networks

Howard Huang
Constantinos B. Papadias
Sivarama Venkatesan

MIMO Communication for Cellular Networks

 Springer

Howard Huang
Bell Labs, Alcatel-Lucent
791 Holmdel Road
Holmdel, New Jersey 07733
USA
howard.huang@alcatel-lucent.com

Constantinos B. Papadias
Athens Information Technology (AIT)
Markopoulo Avenue 19.5 km
190 02 Peania, Athens
Greece
papadias@ait.edu.gr

Sivarama Venkatesan
Bell Labs, Alcatel-Lucent
791 Holmdel Road
Holmdel, New Jersey 07733
USA
venkat.venkatesan@alcatel-lucent.com

ISBN 978-0-387-77521-0   e-ISBN 978-0-387-77523-4
DOI 10.1007/978-0-387-77523-4
Springer New York Dordrecht Heidelberg London

Library of Congress Control Number: 2011942318

Springer is part of Springer Science+Business Media (www.springer.com)

For Michelle, H.H.

For Maria-Anna, C.B.P.

For my parents, S.V.

Preface

Since the 1980s, commercial cellular networks have evolved over several generations, from providing simple voice telephony service to supporting a wide range of other applications (such as text messaging, web browsing, streaming media, social networking, video calling, and machine-to-machine communication) that are accessed through a variety of devices (such as smart phones, laptops, tablet devices, and wireless sensors). These developments have fueled a demand for higher spectral efficiency so that the limited spectral resources allocated for cellular networks can be utilized more effectively.

In parallel, starting in the mid-1990s [1], multiple-input multiple-output (MIMO) wireless communication has emerged as one of the most fertile areas of research in information and communication theory. The fundamental results of this research show that MIMO techniques have enormous potential to improve the spectral efficiency of wireless links and systems. These techniques have already attracted considerable attention in the cellular world, where simple MIMO techniques are already appearing in commercial products and standards, and more sophisticated ones are actively being pursued.

Goals of the book

In this book, we hope to connect these two worlds of MIMO communication theory and cellular network design with the goal of understanding how multiple antennas can best be used to improve the physical-layer performance of a cellular system. We attempt to strike a balance between fundamental theoretical results, practical techniques and core insights regarding the performance limits of multiple antennas in multiuser networks. Unlike books that focus on the theoretical performance of abstract MIMO channels, this one emphasizes the practical performance of realistic MIMO systems.

We present in the first part of the book a systematic description of MIMO capacity and capacity-achieving techniques for different classes of multiple-antenna channels. The second part of the book describes a framework for MIMO system design that accounts for the essential physical-layer features of practical cellular networks. By applying the information-theoretic capacity results to this framework, we present a unified set of system simulation studies that highlight relative performance gains of different MIMO techniques and provides insights into how best to utilize multiple antennas in cellular networks under various conditions. Characterizations of the system-level performance are provided with sufficient generality that the underlying concepts can be applied to a wide range of wireless systems, including those based on cellular standards such as LTE, LTE-Advanced, WiMAX, and WiMAX2.

Intended audience

The book is intended for graduate students, researchers, and practicing engineers interested in the physical-layer design of contemporary wireless systems. The material is presented assuming the reader is comfortable with linear algebra, probability theory, random processes, and basic digital communication theory. Familiarity with wireless communication and information theory is helpful but not required.

Acknowledgements

We have attempted to represent in this book a small sliver of knowledge accumulated by a vast community of researchers. Over the years, we have had the great pleasure of learning from and interacting with many members of this community within Bell Labs, Alcatel-Lucent, and at other corporations and academic institutions. These experts include Angela Alexiou, Alexei Ashikhmin, Dan Avidor, Matthew Baker, Krishna Balachandran, Liyu Cai, Len Cimini, David Goodman, Iñaki Esnaola, Vinko Erceg, Rodolfo Feick, Jerry Foschini, Mike Gans, David Gesbert, Maxime Guillaud, Bert Hochwald, Syed Jafar, Nihar Jindal, Volker Jungnickel, Kemal Karakayali, Achilles Kogiantis, Alex Kuzminskiy, Persa Kyritsi, Angel Lozano, Mike MacDonald, Narayan Mandayam, Laurence Mailaender, Thomas Marzetta, Thomas Michel, Pantelis Monogioudis, Francis Mullany, Marty Meyers, Arogyaswami Paulraj, Farrokh Rashid-Farrokhi, Niranjay Ravindran, Gee Rittenhouse, Dragan Samardzija, James Seymour, Sana Sfar, Steve Simon, Tod Sizer, John Smee, Max Solondz, Robert Soni, Aleksandr Stoylar, Said Tatesh, Stephan ten Brink, Lars Thiele, Filippo Tosato, Cuong Tran, Matteo Trivellato, Giovanni Vannucci, Sergio Verdú, Harish Viswanathan, Sue Walker, Carl Weaver, Thorsten Wild, Stephen Wilkus, Peter Wolniansky, Greg Wright, Gerhard Wunder, Hao Xu, Hongwei Yang, Roy Yates, and Mike Zierdt.

Special thanks go to Antonia Tulino who graciously illuminated various aspects of information theory at a moment's notice. We would also like to specifically acknowledge the following colleagues who provided valuable feedback on an earlier draft: George Alexandropoulos, Federico Boccardi, Dmitry Chizhik, Jonathan Ling, Mohammad Ali Maddah-Ali, Chris Ng, Stelios Papaharalabos, Xiaohu Shang, and Marcos Tavares. We would like to express our gratitude to Francesca Simkin for her keen eye and expert skill in copy editing the manuscript and to Kimberly Howie for providing tips on improving the visual design. Allison Michael and Alex Greene at Springer provided valuable assistance in the production of the manuscript. Finally, we would like to acknowledge our colleague Reinaldo Valenzuela whose enthusiastic leadership has helped shape the spirit and goals of this book.

Howard Huang and Sivarama Venkatesan
Bell Labs, Alcatel-Lucent
Holmdel, New Jersey

Constantinos B. Papadias
Athens Information Technology
Athens, Greece

August 2011

Notation

C^{OL}	SU-MIMO open-loop capacity
$\bar{C}^{\text{OL}}$	SU-MIMO average open-loop capacity
C^{CL}	SU-MIMO closed-loop capacity
$\bar{C}^{\text{CL}}$	SU-MIMO average closed-loop capacity
$\mathcal{C}^{\text{MAC}}$	Multiple-access channel capacity region
C^{MAC}	Multiple-access channel sum-capacity
$\bar{C}^{\text{MAC}}$	Multiple-access channel average sum-capacity
$\mathcal{C}^{\text{BC}}$	Broadcast channel capacity region
C^{BC}	Broadcast channel sum-capacity
$\bar{C}^{\text{BC}}$	Broadcast channel average sum capacity
C^{TDMA}	TDMA channel maximum achievable sum rate
$\bar{C}^{\text{TDMA}}$	TDMA channel average maximum achievable sum rate
M	SU-MIMO channel: number of transmit antennas
	MU-MIMO MAC: number of receive antennas
	MU-MIMO BC: number of transmit antennas
	Cellular system: number of antennas per base
N	SU-MIMO channel: number of receive antennas
	MU-MIMO MAC: number of transmit antennas per user
	MU-MIMO BC: number of receive antennas per user
	Cellular system: number of antennas per user
K	Number of users per base
B	Number of bases
S	Number of sectors per site
$\mathbf{H}, \mathbf{h}$	Complex-valued channel matrix, vector
$\mathbf{s}$	Transmitted signal vector
$\mathbf{Q}$	Transmitted signal covariance
$\mathbf{x}$	Received signal vector
$\mathbf{G}, \mathbf{g}$	Precoding matrix, vector

$\mathbf{n}$	Received noise vector
$\mathbf{u}$	Vector of data symbols
P	Signal power constraint
σ^2	Noise variance
P/σ^2	SU- and MU-MIMO channel: average SNR
	Cellular system: reference SNR
$\lambda_{\max}^2(\mathbf{H})$	Maximum eigenvalue of $\mathbf{HH}^H$
v	Symbol power
q	Quality-of-service weight
α^2	Average channel gain
d	Distance
d_{ref}	Reference distance
Z	Shadow fading realization
G	Directional antenna response
γ	Pathloss coefficient
Γ	Geometry
$\mathbb{E}(x)$	Expected value of random variable x
$\mathbf{A}^H$	Hermitian transpose of matrix $\mathbf{A}$
$\mathrm{tr}\,\mathbf{A}$	Trace of square matrix $\mathbf{A}$
$\mathrm{diag}\,\mathbf{A}$	Diagonal elements of square matrix $\mathbf{A}$
$\mathrm{diag}(a_1,\ldots,a_N)$	Square $N \times N$ matrix with diagonal elements $a_1,\ldots,a_N$
$\mathbf{I}_N$	$N \times N$ identity matrix
$\mathbf{0}_N$	$N \times 1$ vector of zeroes
$\mathbb{C}$	Set of complex numbers
$\mathbb{R}$	Set of real numbers
$\mathcal{B}$	Set of precoding matrices
$\mathcal{U}$	Set of active users

Contents

Chapter 1
Introduction

In a cellular network, a given geographic area is served by several base stations, as shown in Figure 1.1. Each base station (or, simply, *base*) communicates wirelessly using radio-frequency spectrum with one or more mobile terminals (or *users*) assigned to it. Conceptually, it is useful to think of the geographic area as being partitioned into *cells*, with each cell representing the area served by one of the bases. Two-way wireless communication occurs between the users and bases. On the *uplink*, a base detects signals from its assigned users; on the *downlink*, a base transmits signals to its assigned users.

A system engineer's job is to design the cellular network to provide reliable wireless communication using the base station assets over the allocated spectrum resources, where the transmissions are subject to constraints on the radiated power. The network performance measure of most interest to us is the *throughput*, defined as the total data rate transmitted or received by a base station, measured in bits per second (bps). Since we are interested in the performance for a given channel bandwidth, we will focus in this book on the throughput *spectral efficiency* (measured in bps per Hertz) and the spectral efficiency per unit area (measured in bps per Hertz per square kilometer). The emphasis on spectral efficiency is justified by the scarcity and consequent high cost of radio spectrum. Improvements in spectral efficiency can translate into other metrics including higher data rates to the user, improved coverage, improved service reliability, and lower network cost to the operator.

The goal of this book is to explore how the physical-layer performance of contemporary cellular networks can be improved using multiple antennas at the bases and user terminals. Multiple antenna techniques, or *multiple-input multiple-output* (MIMO) techniques, allow us to exploit the *spatial dimen-*

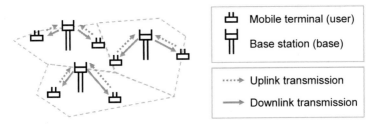

Fig. 1.1 In a cellular network, the geographic area is partitioned into cells, and base stations communicate wirelessly with their assigned users.

sion of the wireless channel, resulting in benefits that include robustness against channel fading, gains in the desired signal power, protection against co-channel interference, and reuse of the spectral resources. MIMO techniques affect the spectral efficiency to varying degrees, and we will investigate the tradeoffs from a system-level perspective between performance gains and implementation complexity for the different techniques.

To begin, we introduce in Section 1.1 some fundamentals of MIMO techniques in the context of isolated channels. In Section 1.2 we give an overview of cellular networks and describe how multiple antennas could be used in this system-level context.

1.1 Overview of MIMO fundamentals

The wireless links in a cellular network can be characterized as either single-user channels or multiuser channels, as shown in Figure 1.2. In this section, we describe some simple models for these channels and give an overview of their theoretical performance limits. These results are described in more detail in Chapters 2 and 3.

1.1.1 MIMO channel models

Single-user MIMO channel
The single-user channel models a point-to-point link between a base and a

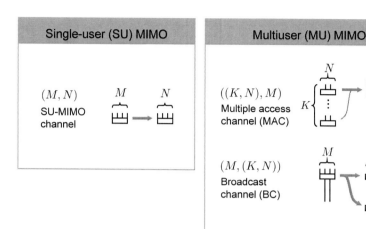

Fig. 1.2 Single-user MIMO communication occurs over a point-to-point link. Multiuser MIMO communication occurs over a multipoint-to-point link (multiple-access channel) or a point-to-multipoint link (broadcast channel).

user. For the downlink, the base is the transmitter and the user is the receiver. For the uplink, the roles are reversed.

We define an (M, N) *single-user* (SU) MIMO channel as a communication link with $M \geq 1$ antennas at the transmitter and $N \geq 1$ antennas at the receiver. (Special cases of the (M, N) MIMO channel are the $(M, 1)$ multiple-input, single-output (MISO) channel, the $(1, N)$ single-input, multiple-output (SIMO) channel, and the $(1,1)$ single-input, single-output (SISO) channel.) The baseband received signal at a given antenna is a linear combination of the M transmitted signals, each modulated by the channel's complex amplitude coefficient, and corrupted by noise. The baseband signal received over a SU-MIMO channel for the duration of a symbol period can be written using vector notation:

$$\mathbf{x} = \mathbf{H}\mathbf{s} + \mathbf{n}, \tag{1.1}$$

where

- $\mathbf{x} \in \mathbb{C}^{N \times 1}$ is the received signal whose nth element $(n = 1, \ldots, N)$ is associated with antenna n
- $\mathbf{H} \in \mathbb{C}^{N \times M}$ is the channel matrix whose (n, m)th entry $(n = 1, \ldots, N; m = 1, \ldots, M)$ gives the complex amplitude between the mth transmit antenna and the nth receive antenna

- $\mathbf{s} \in \mathbb{C}^{M \times 1}$ is the transmitted signal vector, having covariance $\mathbf{Q} := \mathbb{E}\left(\mathbf{s}\mathbf{s}^H\right)$, and subject to the power constraint $\operatorname{tr} \mathbf{Q} \leq P$
- $\mathbf{n} \in \mathbb{C}^{N \times 1}$ is a circularly symmetric complex Gaussian vector representing additive receiver noise, with mean $\mathbb{E}(\mathbf{n}) = \mathbf{0}_N$ and covariance $\mathbb{E}(\mathbf{n}\mathbf{n}^H) = \sigma^2 \mathbf{I}_N$

In practice, the wireless channel experiences fading in both the frequency and time domains. However, for simplicity we assume that the channel bandwidth is small compared to the coherence bandwidth so that the channel is frequency-nonselective. In environments where the transmitter and receiver are stationary, we can model $\mathbf{H}$ as being fixed for the duration of a coding block (on the order of hundreds or thousands of symbol periods) but changing randomly from one block to another. If the antennas at the transmitter and receiver are spaced sufficiently far apart (with respect to the channel angle spread), the elements of $\mathbf{H}$ can be modeled as realizations of independent and identically distributed (i.i.d.) Gaussian random variables with zero mean and unit variance. Because the amplitude of each element has a Rayleigh distribution, this channel is said to be drawn from an *i.i.d. Rayleigh* distribution. Because the channel coefficients of $\mathbf{H}$ are independent and therefore uncorrelated, the channel is said to be *spatially rich*. If the elements of $\mathbf{H}$ have unit variance, then the average signal-to-noise ratio (SNR) at any of the N receive antennas is P/σ^2.

Multiuser MIMO channels

We consider two types of multiuser channels: the *multiple-access channel* (MAC) and the *broadcast channel* (BC).

The MAC is used to model a single base receiving signals from multiple users on the uplink of a cellular network. We will use the notation $((K, N), M)$ to denote a MAC with $K \geq 1$ users, each with $N \geq 1$ antennas, whose signals are received by a base with $M \geq 1$ antennas. We use the term *MIMO MAC* to denote a MAC with multiple ($M > 1$) base antennas serving multiple ($K > 1$) users, where each user has one or more antennas.

The data signals sent by the users are independent, and the base receives the sum of K signals modulated by each user's MIMO channel and corrupted by noise. The baseband received signal can be written as:

$$\mathbf{x} = \sum_{k=1}^{K} \mathbf{H}_k \mathbf{s}_k + \mathbf{n}, \tag{1.2}$$

where

- $\mathbf{x} \in \mathbb{C}^{M \times 1}$ is the received signal whose mth element $(m = 1, \ldots, M)$ is associated with antenna m
- $\mathbf{H}_k \in \mathbb{C}^{M \times N}$ is the channel matrix for the kth user $(k = 1, \ldots, K)$ whose (m, n)th entry $(m = 1, \ldots, M; n = 1, \ldots, N)$ gives the complex ampli- tude between the nth transmit antenna of user k and the mth receive antenna. Each coefficient for the kth user's channel $\mathbf{H}_k$ is assumed to be i.i.d. Rayleigh with unit variance, and the channels are mutually indepen- dent among the users
- $\mathbf{s}_k \in \mathbb{C}^{N \times 1}$ is the transmitted signal vector from user k. The covariance is $\mathbf{Q}_k := \mathbb{E}\left(\mathbf{s}_k \mathbf{s}_k^H\right)$, and the signal is subject to the power constraint $\operatorname{tr} \mathbf{Q}_k \leq P_k$
- $\mathbf{n} \in \mathbb{C}^{M \times 1}$ is a circularly symmetric complex Gaussian vector representing additive receiver noise, with mean $\mathbb{E}(\mathbf{n}) = \mathbf{0}_M$ and covariance $\mathbb{E}(\mathbf{n}\mathbf{n}^H) = \sigma^2 \mathbf{I}_M$

The BC is used to model a single base serving multiple users in the down- link of a cellular network. Each user receives independent data, so the infor- mation is not broadcast in the sense of users receiving the same data. We use the shorthand $(M, (K, N))$ to denote a base station with $M \geq 1$ transmit antennas serving $K \geq 1$ users, each with $N \geq 1$ antennas. We use the term *MIMO BC* to denote a BC with multiple $(M > 1)$ base antennas serving multiple $(K > 1)$ users, where each user has one or more antennas. The base transmits a common signal $\mathbf{s}$, which contains the encoded symbols of the K users' data streams. This signal travels over the channel $\mathbf{H}_k^H$ to reach user k $(k = 1, \ldots, K)$. (The Hermitian transpose allows for convenient comparisons between the MAC and BC, as we will see later in Chapter 3.) The baseband received signal by the kth user can be written as:

$$\mathbf{x}_k = \mathbf{H}_k^H \mathbf{s} + \mathbf{n}_k, \tag{1.3}$$

where

- $\mathbf{x}_k \in \mathbb{C}^{N \times 1}$ whose nth element $(n = 1, \ldots, N)$ is associated with antenna n of user k
- $\mathbf{H}_k^H \in \mathbb{C}^{N \times M}$ is the channel matrix for the kth user $(k = 1, \ldots, K)$ whose (n, m)th entry $(n = 1, \ldots, N; m = 1, \ldots, M)$ gives the complex amplitude between the nth transmit antenna of user k and the mth receive antenna. Each coefficient for the kth user's channel $\mathbf{H}_k^H$ is assumed to be i.i.d.

Rayleigh with unit variance, and the channels are mutually independent among the users

- $\mathbf{s} \in \mathbb{C}^{M \times 1}$ is the transmitted signal vector, which is a function of the data signals for the K users. The covariance is $\mathbf{Q} := \mathbb{E}\left(\mathbf{s}\mathbf{s}^H\right)$, and the signal is subject to the power constraint $\operatorname{tr}\mathbf{Q} \leq P$
- $\mathbf{n}_k \in \mathbb{C}^{N \times 1}$ is a circularly symmetric complex Gaussian vector representing additive receiver noise, with mean $\mathbb{E}(\mathbf{n}) = \mathbf{0}_N$ and covariance $\mathbb{E}(\mathbf{n}\mathbf{n}^H) = \sigma^2\mathbf{I}_N$

1.1.2 Single-user capacity metrics

For a single-user channel, the performance metric of interest is the *spectral efficiency*, defined as the data rate achieved per unit of bandwidth and measured in units of bits per second per Hertz. When it is understood that the bandwidth is held fixed, we will often use *rate* and *spectral efficiency* interchangeably. From information theory, the *Shannon capacity* (or simply, capacity) of the SU-MIMO channel is the maximum spectral efficiency at which reliable communication is possible, meaning that the bit error rate can be made arbitrarily small with sufficiently long coding blocks over many symbol periods. Here we describe the capacity of the SISO, SIMO, MISO, and MIMO channels when the channels are time-invariant. Rather than use information theory to derive the capacity, we instead focus on the spectral efficiency of techniques for approaching the capacity limits in practice.

1.1.2.1 SISO capacity

From (1.1), the received signal over a SISO channel can be written as

$$x = hs + n, \tag{1.4}$$

where h is the fixed scalar complex amplitude of the channel. A sufficient statistic for detecting s is obtained by multiplying the received signal by the complex conjugate of h to yield

$$h^*x = |h|^2 s + h^*n. \tag{1.5}$$

The sufficient statistic has an effective SNR of $|h|^2 P/\sigma^2$, and the capacity of this channel is

$$C = \log_2 \left(1 + \frac{P|h|^2}{\sigma^2} \right) \text{ bps/Hz}. \tag{1.6}$$

In Gaussian channels, one can achieve spectral efficiency very close to capacity using contemporary codes such as turbo codes or low-density parity check (LDPC) codes in combination with iterative decoding algorithms. We will loosely refer to such codes as capacity-achieving (although near-capacity-achieving would be more accurate). Since (1.5) is equivalent to a Gaussian channel, we can (nearly) achieve capacity using these optimal codes if the channel h is known ideally at the receiver to provide the sufficient statistic.

1.1.2.2 SIMO capacity

We now consider a $(1, N)$ SIMO channel, with N antennas at the receiver:

$$\mathbf{x} = \mathbf{h}s + \mathbf{n}, \tag{1.7}$$

where $\mathbf{h} \in \mathbb{C}^{N \times 1}$ is the time-invariant SIMO channel. A sufficient statistic is obtained by taking the inner product of the received signal $\mathbf{x}$ with the channel $\mathbf{h}$:

$$\mathbf{h}^H \mathbf{x} = \|\mathbf{h}\|^2 s + \mathbf{h}^H \mathbf{n}. \tag{1.8}$$

In doing so, the SIMO channel is reduced to a scalar Gaussian channel. This linear receiver combining maximizes the output SNR, and it is known as maximal ratio combining (MRC). The SNR is $\|\mathbf{h}\|^2 P/\sigma^2$, and the capacity is

$$C = \log_2 \left(1 + \frac{P\|\mathbf{h}\|^2}{\sigma^2} \right) \text{ bps/Hz}. \tag{1.9}$$

The capacity of the $(1, N)$ SIMO channel can be achieved using optimal capacity-achieving codes and a linear combiner $\mathbf{h}^H$ as shown in the top half of Figure 1.3. As a result of combining, the multiple receive antennas provide a *power gain* in the output SNR. For example, if the absolute value of each channel coefficient is 1 ($|h_n| = 1$ for $n = 1, \ldots, N$), then $\|\mathbf{h}\|^2 = N$, and the N antennas provide a power gain of N compared to the SISO channel. Similarly, if the channel elements are i.i.d. Rayleigh (with unit mean variance), then the average power gain is $\mathbb{E}(\|\mathbf{h}\|^2) = N$.

The impact of the power gain on the capacity (1.9) depends on the channel SNR P/σ^2. We can use the following expressions to approximate the capacity performance for very high and very low SNRs:

$$\log_2(1+x) \approx \log_2(x) \quad \text{if } x \gg 1 \tag{1.10}$$

$$\log_2(1+x) \approx x\log_2(e) \quad \text{if } x \ll 1. \tag{1.11}$$

For the case where the channel SNR P/σ^2 is very high, we see from (1.10) that doubling the output SNR due to combining results in a 1 bps/Hz increase in the capacity. In other words, if the output SNR increases linearly, the capacity increases logarithmically. On the other hand if the channel SNR is very low, we see from (1.11) that a linear increase in the output SNR results in a linear increase in the capacity. Therefore multiple antennas in the SIMO channel boost the capacity more significantly at low SNRs.

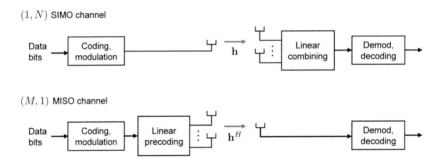

Fig. 1.3 Capacity is achieved over a SIMO channel with linear combining matched to the channel $\mathbf{h} \in \mathbb{C}^{N \times 1}$. Capacity is achieved over a MISO channel with a linear precoder matched to the channel $\mathbf{h} \in \mathbb{C}^{1 \times M}$.

1.1.2.3 MISO capacity

For an $(M, 1)$ MISO channel ($\mathbf{h} \in \mathbb{C}^{1 \times M}$), where $\mathbf{h}$ is known at the transmitter, the capacity is

$$C = \log_2\left(1 + \frac{P\|\mathbf{h}\|^2}{\sigma^2}\right) \text{ bps/Hz.} \tag{1.12}$$

This capacity can be achieved using the structure shown in bottom half of Figure 1.3. The transmitted signal at antenna m ($m = 1, \ldots, M$) is $s_m =$

$g_m u$, where g_m is a complex-valued weight applied to an encoded data symbol u. Weighting the data symbol in this manner at the transmitter is known as linear precoding, or simply *precoding*. To achieve the MISO capacity, the precoding weights are chosen to match the channel $\mathbf{h}$ so that the weight on the mth antennas is the normalized complex conjugate of the mth antenna's channel: $g_m = h_m^*/\|\mathbf{h}\|$. If the data symbols have power $\mathbb{E}(|u|^2) = P$, then the transmitted signal also has power P because $\operatorname{tr}\mathbb{E}(\mathbf{ss}^H) = \operatorname{tr}\mathbb{E}(\mathbf{g}uu^*\mathbf{g}^H) = P$. The received signal is

$$x = \mathbf{h}\mathbf{s} + n = \|\mathbf{h}\|\, u + n, \tag{1.13}$$

and the SNR is $\|\mathbf{h}\|^2 P/\sigma^2$, yielding the capacity (1.12). This capacity is the same as the SIMO channel's (1.9), implying that the power gain achieved using combining over the SIMO channel can be achieved using precoding over the MISO channel if the channel is known at the transmitter.

Using precoding with weights matched to the MISO channel is similar to MRC matched to the SIMO channel. The channel-matched precoding is sometimes known as *maximal ratio transmission* (MRT) because it maximizes the received SNR. If the channel is line-of-sight (or has a very small angle spread) and if the transmit antennas are closely spaced (by about half the carrier wavelength), MRT creates a directional beam pointed towards the receiver. This type of precoding is sometimes known as transmitter beamforming.

1.1.2.4 SU-MIMO capacity

If multiple antennas are used at both the transmitter and receiver, the (M, N) MIMO channel could support multiple data streams. The data streams are *spatially multiplexed* and are sent simultaneously over the same frequency resources across the M antennas. The receiver uses N antennas to demodulate the streams based on their spatial characteristics. The capacity of the MIMO channel depends on knowledge of the channel state information (CSI) $\mathbf{H}$. For now, we consider the *closed-loop* MIMO capacity that assume CSI is known at both the transmitter and receiver. (The *open-loop* MIMO capacity assumes that CSI is known only at the receiver and will be discussed in Chapter 2).

If both the transmitter and receiver have ideal knowledge of CSI, the MIMO channel can be decomposed into $r \geq 1$ parallel (non-interfering) scalar Gaussian subchannels, where r is the rank of the MIMO channel $\mathbf{H}$. For simplicity, we assume here that $\mathbf{H}$ is full rank so that $r = \min(M, N)$. This decomposition is achieved by applying a precoding matrix at the transmitter and a combining matrix at the receiver. These matrices are derived from the eigendecomposition of $\mathbf{H}^H \mathbf{H}$ and $\mathbf{H} \mathbf{H}^H$, respectively, and therefore the resulting parallel subchannels are often referred to as *eigenmodes*.

The SNRs of the r scalar Gaussian subchannels turn out to be proportional to the eigenvalues $\lambda_1^2, \lambda_2^2, \ldots, \lambda_r^2$ of $\mathbf{H} \mathbf{H}^H$ (or $\mathbf{H}^H \mathbf{H}$). More precisely, if the transmitter allocates power P_j to subchannel j, then the SNR achieved on it equals $\lambda_j^2 P_j / \sigma^2$ and the corresponding data rate is

$$\log_2(1 + \lambda_j^2 P_j / \sigma^2). \tag{1.14}$$

The capacity of the composite MIMO channel can then be expressed as the sum of the r subchannel capacities (1.14), maximized over all power allocations $P_1, P_2, \ldots, P_r$ subject to the constraint $\sum_{j=1}^r P_j \leq P$. The resulting (closed-loop) MIMO capacity is

$$C^{\mathrm{CL}}(\mathbf{H}, P/\sigma^2) = \max_{P_j, \sum P_j \leq P} \sum_{j=1}^{r} \log_2 \left(1 + \frac{\lambda_j^2 P_j}{\sigma^2}\right) \text{ bps/Hz}. \tag{1.15}$$

The optimal power allocation depends on the SNR P/σ^2 and can be solved using an algorithm known as *waterfilling*. We note that the SIMO capacity (1.9) and MISO capacity (1.12) are special cases of (1.15) where $r = 1$.

The MIMO capacity (1.15) can be achieved using the technique shown in Figure 1.4. The data stream is multiplexed into r substreams, and each substream is encoded with a capacity-achieving code corresponding to its maximum achievable data rate (1.14). These streams are precoded and transmitted over M antennas. At the receiver, the combiner is applied to yield r substreams signals. These are independently demodulated and decoded, and the estimated data bits are demultiplexed to reconstruct the original information data stream.

It can be shown that at very low SNRs, it is optimal according to the waterfilling algorithm to transmit with full power P on the dominant eigenmode, which is the one corresponding to the largest eigenvalue $\max_j \lambda_j^2$. In other words, power is transmitted on only one of the r subchannels. From (1.15) and (1.11), the resulting capacity is approximately

$$\log_2 \left(1 + \max_j \lambda_j^2 \frac{P}{\sigma^2}\right) \approx \max_j \lambda_j^2 \frac{P}{\sigma^2} \log_2 e. \tag{1.16}$$

The largest eigenvalue value reflects the amplitude boost on the dominant eigenmode due to precoding and combining. Therefore increasing the number of antennas increases the capacity through the resulting power gain.

At very high SNRs, it is optimal to allocate equal power P/r on each of the r subchannels. Using (1.15), the resulting capacity is approximately

$$\sum_{j=1}^{r} \log_2 \left(1 + \frac{\lambda_j^2 P}{r\sigma^2} \right) \approx \min(M, N) \log_2 \frac{P}{\sigma^2}, \tag{1.17}$$

which follows from (1.10). The factor $\min(M, N)$ in (1.17) is known as the *multiplexing gain* and indicates the number of interference-free subchannels resulting from the decomposition of **H**. It is also known as the *pre-log factor* and the *spatial degrees of freedom*.

As a result of the multiplexing gain, the capacity increases linearly as the number of antennas increases. Therefore for a fixed transmit power and bandwidth at high SNR, doubling the number of transmit and receive antennas results in a doubling of the capacity. In order to achieve higher data rates using fixed bandwidth resources and without using additional antennas (for example, over a SISO channel), the transmit power needs to increase exponentially to support a linear gain in capacity at high SNRs (1.6). This solution would be impractical due to the prohibitive cost of larger amplifiers, and possibly unlawful with regard to transmission regulations.

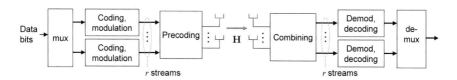

Fig. 1.4 The closed-loop MIMO capacity can be achieved by decomposing the SU-MIMO channel **H** into $r = \min(M, N)$ parallel subchannels (eigenmodes) using appropriate precoding and combining matrices.

1.1.3 Multiuser capacity metrics

For single-user MIMO channels, reliable communication can be achieved for any rate less than the capacity C. For a K-user multiuser MIMO channel, the appropriate performance limit is the K-dimensional *capacity region* consisting of all the vectors of rates $\mathbf{R} := (R_1, \dots, R_K)$ that can be achieved *simultaneously* by the K users (here, $R_k \geq 0$ is the rate corresponding to user k). The MAC capacity region $\mathcal{C}^{\mathrm{MAC}}$ is a function of the users' channels $\mathbf{H}_k$, $k = 1, \dots, K$ and SNRs P_k/σ^2, $k = 1, \dots, K$:

$$\mathcal{C}^{\mathrm{MAC}} \left(\mathbf{H}_1, \dots, \mathbf{H}_K, P_1/\sigma^2, \dots, P_K/\sigma^2 \right), \tag{1.18}$$

and the BC capacity region is a function of the channels $\mathbf{H}_k^H$, $k = 1, \dots, K$ and the SNR P/σ^2:

$$\mathcal{C}^{\mathrm{BC}} \left(\mathbf{H}_1^H, \dots, \mathbf{H}_K^H, P/\sigma^2 \right). \tag{1.19}$$

To illustrate the characteristics of the multiuser capacity metrics, we consider the generic two-user capacity regions shown in Figure 1.5. In general, the rate regions for the MAC and BC are convex regions. Along the boundary of the regions in Figure 1.5, it is not possible to increase the rates of both users simultaneously because as the rate of one user increases, the rate of the other user must decrease. This tradeoff occurs because the two users must share common resources.

Given the capacity region for a K-user MU-MIMO channel, it is useful to have a scalar metric that indicates a single "best" rate vector belonging to the capacity region. For example, assuming that all users pay the same per bit of information communicated, the highest revenue would be achieved by maximizing the sum of the rates delivered to all the users. This *sum-rate capacity* (or *sum-capacity*) metric is defined as:

$$\max_{\mathbf{R} \in \mathcal{C}} \sum_{k=1}^{K} R_k, \tag{1.20}$$

where $\mathcal{C}$ denotes either the BC or MAC capacity region. From the perspective of a cellular base station, the sum-capacity is useful because it measures the maximum downlink throughput transmitted by the base over a BC or the maximum uplink throughput over a MAC.

In Figure 1.5, the sum-capacity is achieved by the rate vector corresponding to the point C on the capacity region boundary at which the tangent line has slope -1.

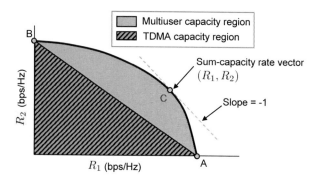

Fig. 1.5 Generic 2-user capacity region (MAC or BC). Sum-capacity is achieved at the point C on the boundary. Also shown is the TDMA rate region, corresponding to time-multiplexing the users.

Multiple-access channel capacity region

Suppose we regard Figure 1.5 as representing the multiple-access channel capacity region. Then point A corresponds to having user 1 alone transmit at maximum power and having user 2 be silent (and vice versa for point B). Any rate vector lying on the line segment joining A and B can be achieved by time-multiplexing the transmissions of user 1 and user 2 (with each point on that line corresponding to a different time split between them). The triangular region bounded by the axes and the line segment joining A and B is therefore known as the time-division multiple-access (TDMA) rate region.

In general, there are rate vectors belonging to the MAC capacity region that lie outside the TDMA rate region, and these are achieved through spatial multiplexing by having the users transmit *simultaneously* over the same frequency resources and jointly decoding them. The spatial multiplexing of signals from multiple users is known as *space-division multiple-access* (SDMA). For the case of a single mobile antenna ($N = 1$), the multiplexing of users is implemented by having each user transmit independently encoded and modulated data signals simultaneously, as shown in the top half of Figure 1.6.

Unlike the closed-loop strategy where precoding and combining creates parallel subchannels, each user's signal is received in the presence of interference from the other users. To account for the interference optimally, the receiver uses *multiuser detection* to jointly detect the spatially multiplexed signals. Specifically, the receiver is based on a minimum mean-squared error (MMSE) combiner followed by a *successive interference canceller* (SIC). The SIC decodes and cancels substreams in an ordered manner so each user's signal is received in the presence of interference from those only the users yet to be decoded. The MMSE-SIC architecture is more complex than the combiner and independent decoders used for achieving closed-loop MIMO capacity (Figure 1.4), but this is the price we pay for achieving spatial multiplexing if the transmitting antennas cannot cooperate.

At very low SNRs, the interference power is dominated by the thermal noise power. In this regime, using an MRC is optimal, and the capacity achieved by the kth user, averaged over i.i.d. Rayleigh realizations of the channel $\mathbf{h}_k$, is approximately

$$N \frac{P_k}{\sigma^2} \log_2 e, \qquad (1.21)$$

where $N = \mathbb{E}(\|\mathbf{h}\|^2)$ represents the average combining gain achieved from the multiple receive antennas. To maximize the achievable sum rate, each user transmits with full power (P_k for user k), and the resulting average sum-capacity at very low SNR is

$$N \frac{P}{\sigma^2} \log_2 e, \qquad (1.22)$$

where we use P to denote the total power of the K users: $P = \sum_{k=1}^{K} P_k$. The sum-capacity does not depend on how the power is distributed among the users or even how many users there are, as long as the total power is fixed. Therefore at low SNR, the N antennas provide a power gain, but the sum-capacity is independent of the number of users K for a fixed total power P.

At very high SNRs, the sum-capacity of the MAC is approximately

$$\min(K, M) \log_2 \frac{P}{\sigma^2}, \qquad (1.23)$$

implying that $\min(K, M)$ interference-free links can be established over the MAC. This capacity can be achieved using the MMSE-SIC to disentangle the users' signals. The multiplexing gain of $\min(K, M)$ in (1.23) is the same as that of a (K, M) SU-MIMO link with CSI and precoding at the transmitter (1.17). It follows that full multiplexing gain could have been achieved over the SU-MIMO link with an MMSE-SIC and without CSI and transmitter precoding.

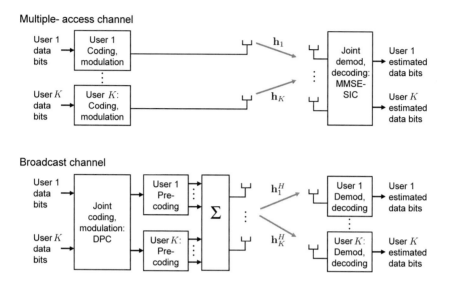

Fig. 1.6 Strategies for achieving rate vectors on the capacity region boundary ($N = 1$). For the MAC, users transmit signals simultaneously and are jointly decoded using MMSE-SIC. For the BC, users' data streams are jointly encoded using dirty paper coding and then precoded.

Broadcast channel capacity region

If we now regard Figure 1.5 as the broadcast channel capacity region, then point A (resp. B) corresponds to transmitting exclusively to user 1 (resp. user 2) at full power. Again, points on the line segment joining A and B are achievable by time-sharing the channel between the users, and the triangular region bounded by the axes and the A-B line segment is called the time-division multiple-access (TDMA) rate region.

As was the case for the MAC, rate vectors on the capacity region boundary of the BC are achieved by spatially multiplexing signals for multiple users. One method for multiplexing is linear precoding as shown in Figure 1.7, where the users' data streams are independently encoded and transmitted simultaneously using different linear precoding vectors. The precoding vectors are a function of the channel realizations $\mathbf{h}_1^H, \ldots, \mathbf{h}_K^H$, and hence CSI is required at the transmitter. For example, MRT precoding could be implemented for each user. (If the antennas are highly correlated, each MRT vector creates a directional beam pattern pointing in the direction of its desired user as shown in Figure 1.7.) Because the MRT vector is determined myopically for each user and does not account for the interaction between beams, each user would experience interference from signals intended for the other users.

In general, SDMA implemented with linear precoding alone is not optimal in that it does not achieve rate vectors lying on the BC capacity region boundary. The optimal technique uses linear precoding preceded by a joint encoding and modulation technique known as *dirty paper coding* (DPC). The DPC encoder output provides data symbols for each of the K users which are precoded and transmitted simultaneously. DPC uses knowledge of the users' channels and data streams to perform coding in an ordered fashion among the users, and thereby to remove interference at the transmitter. This is similar to the effect of the SIC which removes the interference at the receiver through ordered decoding. The DPC and precoders are designed so that each user's received signal does not experience interference from users encoded after it.

At very low SNRs, the BC capacity region becomes equivalent to the corresponding TDMA capacity region, and SDMA no longer provides any benefit over simple single-user transmission. In this regime, the sum rate is maximized by serving the single user whose MISO channel yields the highest capacity. From (1.12) and (1.11), the resulting sum-capacity at asymptotically low SNRs is approximately

$$\max_{k=1,\ldots,K} \|\mathbf{h}_k\|^2 \frac{P}{\sigma^2} \log_2 e. \tag{1.24}$$

Therefore the multiple transmit antennas provide power gain through precoding, and additional users increase the capacity according to the statistics of $\max_{k=1,\ldots,K} |\mathbf{h}_k|^2$.

At high SNRs, the sum-capacity of the BC is the same as the MAC's (1.23) if the BC power P is the same as the total power of the MAC users:

$$\min(K, M) \log_2 \frac{P}{\sigma^2}. \tag{1.25}$$

The multiplexing gain $\min(K, M)$ achieved over the BC requires CSI at the transmitter but does not require coordination between the users. Therefore a

(K, M) SU-MIMO channel could achieve the same multiplexing gain without jointly processing the received signals across the antennas.

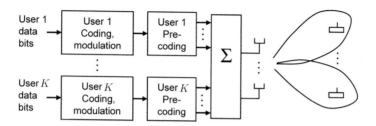

Fig. 1.7 Suboptimal spatial multiplexing for the BC can be implemented as spatial division multiple-access (SDMA) by simultaneously transmitting multiple precoded data streams. If the antennas are closely spaced and highly correlated, precoding creates directional beams.

1.1.4 MIMO performance gains

To illustrate the performance gains of the multiple antenna techniques, we compare the capacity performance over the following channels:

- (1,1) SU-SISO with channel $h \in \mathbb{C}$ and power constraint P
- (4,4) SU-MIMO with channel $\mathbf{H} \in \mathbb{C}^{4 \times 4}$ and power constraint P
- ((4,1),4) MU MAC with channel $\mathbf{h}_k \in \mathbb{C}^{4 \times 1}$ and power constraint $P_k = P/4$ for each user $k = 1, \ldots, 4$
- (4,(4,1)) MU BC with channel $\mathbf{h}_k^H \in \mathbb{C}^{1 \times 4}$ for each user $k = 1, \ldots, 4$ and power constraint P

The channels are assumed to all be i.i.d. Rayleigh, and the noise is assumed to have variance σ^2 so the average SNR in all cases is P/σ^2. For the single-user channels, we consider the SISO capacity (1.6) and closed-loop MIMO capacity (1.15) averaged over the i.i.d. channel realizations. For the multiuser channels, we consider the sum-capacity averaged over the i.i.d. channel realizations.

To better observe the multiplexing gains at high SNR and power gains at low SNR, the capacity performances are illustrated in Figures 1.8 and 1.9 for the range of high and low SNRs, respectively.

Capacity at high SNR

At high SNRs, multiple antennas provide capacity gains through spatial multiplexing. For the SU channels, the multiplexing gain $\min(M, N)$ (1.17) gives the slope of the capacity with respect to $\log_2 P$ for asymptotically high SNR. The capacity curves for SISO and MIMO are parallel, respectively, to the lines given by $\log_2(P/\sigma^2)$ and $4\log_2(P/\sigma^2)$. For every increase of SNR by 3 dB, the MIMO capacity increases by 4 bps/Hz and the SISO capacity increases by 1 bps/Hz. While the scaling in (1.17) applies to asymptotically high SNR, the asymptotic behavior is already apparent in Figure 1.8 for an SNR of 15 dB. Increasing the number of antennas results in a linear capacity increase, whereas increasing the transmit power results in only a logarithmic increase. Therefore achieving very high data rates using SISO becomes infeasible due to the required power. For example, at $P/\sigma^2 = 15$ dB, the MIMO capacity is about 16 bps/Hz. Achieving the same capacity with SISO requires an SNR of 48 dB, so MIMO saves more than three orders of magnitude of power. The significant capacity gains and power savings provided by spatial multiplexing are the key benefits that make MIMO techniques so appealing at high SNR.

We can also compare the SU-MIMO channel performance with the BC and MAC performance. Because the multiplexing gain of the MU-MIMO channels is $\min(K, M) = 4$, the sum-capacity slope with respect to SNR is the same as the SU-MIMO channel. Figure 1.8 shows that the absolute capacities (and not just their slopes) are nearly identical, indicating that at high SNR, little is gained by coordinating the transmission or reception among multiple single-antenna users.

Capacity at low SNR

For the SU-MIMO channel at low SNR, transmitting a single stream is optimal. In this regime, a factor of α power gain in the SNR results in a capacity gain of α. In some cases, these gains can be larger than those attained from multiplexing at high SNR. For example, as highlighted in Figure 1.9, at SNR = -15 dB, the power gain achieved from precoding and receiver combining over the (4,4) SU-MIMO channel results in a factor of 9 gain in capacity compared to SU-SISO. Even though it is no longer optimal to transmit a single stream at higher SNRs, the capacity gain of MIMO versus SISO is greater than 4 in the range of SNRs shown in this figure.

At low SNRs there is a significant difference between the capacity of the SU-MIMO channel and the sum-capacity of either the MAC or BC. It follows that in the regime of low SNR, coordination at both the transmitter and receiver, respectively in the form of precoding and combining, provides benefit over having coordination at just one end or the other.

1.2 Overview of cellular networks

The previous section describes the capacity performance of narrowband MIMO channels that model the communication links in a cell between a single base and its assigned user(s). In a cellular network consisting of multiple bases and multiple assigned users per base, the transmissions from one cell may cause co-channel interference at the receivers of other cells, resulting in degraded performance. MIMO links form the basis of the cellular network's underlying physical layer, and in order to efficiently leverage the benefits of multiple antennas, it is necessary to understand the system-level aspects of the network architecture, the nature of co-channel interference, and how the interference can be mitigated [2].

1.2.1 System characteristics

In this section we describe the system-level aspects of cellular networks that are relevant for modeling physical-layer performance. These include the partitioning of the geographic region into cells and sectors, the technique for supporting multiple users in the network, and the allocation of spectral resources in packet-based systems.

Cell sites and sectorization

In practice, the location of base stations depends on factors such as the user distribution, terrain, and zoning restrictions. However for the purpose of simulations, it is convenient to assume a hexagonal grid of cells where a base is located at the center of each cell, as shown in Figure 1.10. Each user is assigned to the base with the best radio link quality which, as a result of channel shadow fading, may not necessarily be the one that is closest to it. A user lying in a particular hexagonal cell is not necessarily assigned to the

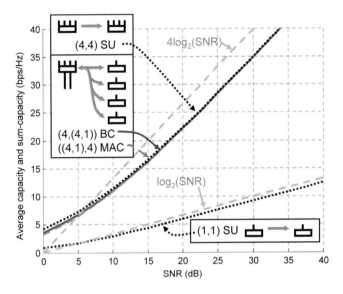

Fig. 1.8 At high SNR, the average (M, N) SU-MIMO capacity scales linearly with the number of antennas $\min(M, N)$ as a result of spatially multiplexing independent data streams. The multiplexing gain gives the slope of the capacity with respect to $\log_2(P/\sigma^2)$ at high SNR.

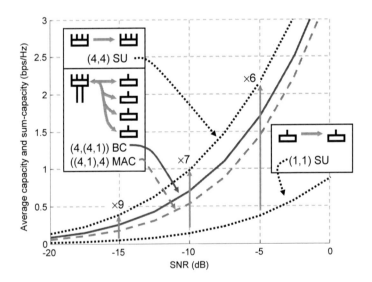

Fig. 1.9 At low SNR, the SU-MIMO capacity gains over SU-SISO are greater compared to the gains in the high-SNR regime. Coordinating both the transmit and receive antennas (as in the SU-MIMO channel) provides benefits over coordination at only one end or the other.

base at the center of that cell. The cell boundaries therefore highlight the location of the bases but do not indicate the assigned bases.

A fundamental characteristic of cellular networks is that the spectral resources are reused at different cells in order to improve the spectral efficiency. In general, the channel bandwidth is partitioned into subbands and these are assigned to different cells to trade off between the harmful effects of interference and the gains from frequency reuse. For example in early analog cellular networks, the network was partitioned into groups of at least seven cells, and each cell within a cluster was assigned a set of subbands which did not interfere with those assigned to other cells in the cluster. This concept of *reduced frequency reuse* is shown in the left subfigure of Figure 1.11. A user assigned to a particular cell received interference from cells in adjacent clusters, but the interference power was low enough to ensure reliable demodulation of the desired signal. With the introduction of digital communication techniques, reliable demodulation could be achieved in the presence of higher interference power. In contemporary cellular networks, the entire channel bandwidth is reused at each cell under *universal frequency reuse*. Compared to a system with reduced frequency reuse, users assigned to a particular cell experience more interference, but each cell uses more bandwidth so the overall system spectral efficiency is potentially higher.

At each cell site, the spatial characteristics of the channels are determined by the base station antenna architecture and the transceiver technique. Figure 1.12 shows five examples of different architectures. Column A shows a single omni-directional antenna for the base. An omni-directional antenna can be implemented as a dipole element that has a circular cross-section. Multiple antenna techniques could be implemented using three omni-directional antennas at the base, as shown in Column B.

Multiple antennas at each site could also be used for *sectorization*, which is the radially partitioning of a cell site into multiple spatial channels. In practice, cell sites are often partitioned into three sectors, as shown in the right subfigure of Figure 1.10 and implemented using the antenna architecture in column C of Figure 1.12. Each antenna element has a rectangular cross-section and uses a physical reflector to focus energy in a beam towards the desired direction. Sectorization is an efficient technique for achieving higher spectral efficiency if the bandwidth is reused at each sector and if the interference between sectors is tolerable. MIMO techniques can be implemented in a sector by deploying multiple directional antennas, as shown in column D. If the antennas are closely spaced, directional beams (Figure 1.7) can be formed from each array. In column E, two closely-spaced directional antennas are deployed in each sector, and these antenna pairs each form two directional beams. From the perspective of the user, each beam is indistinguishable from the antenna pattern of a sector served by a single directional antenna. Therefore a total of six sectors, each with $M = 1$ effective antenna, can be formed at the site using this antenna configuration.

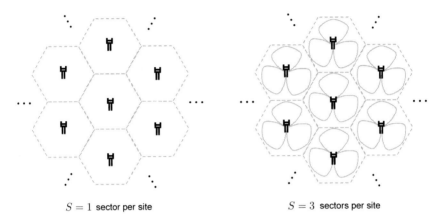

$S = 1$ sector per site $S = 3$ sectors per site

Fig. 1.10 A hexagonal grid of cells will be used for performance simulations. On the left, the antennas at each base are omni-directional. On the right, each cell site is partitioned radially into three sectors.

Multiple access

In order to support multiple users in a cellular network, a cellular standard defines a multiple-access technique to specify how the spectral resources are shared among the users assigned to a cell or sector. These techniques include frequency division multiple-access (FDMA), time division multiple-access (TDMA), code division multiple-access (CDMA), and orthogonal frequency division multiple-access (OFDMA) [3].

In first-generation analog cellular networks based on FDMA, the channel bandwidth is partitioned into subcarriers, and a user is assigned to a single subcarrier. Subcarriers can be reused at different cells if they are sufficiently far apart that the interference power is negligible. TDMA builds on FDMA by imposing a slotted time structure on each subchannel. By assigning users to different time slots, multiple users can share each subcarrier, thereby increas-

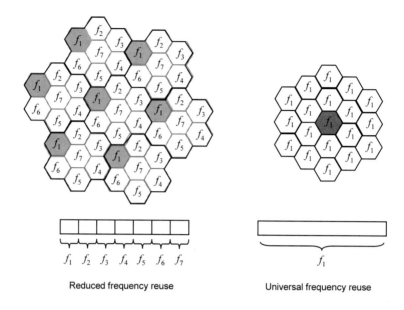

Reduced frequency reuse　　　　　　　　　　Universal frequency reuse

Fig. 1.11 Under universal frequency reuse, cells operate on the same frequency. Under reduced frequency reuse, cells operate on a fixed reuse pattern to mitigate interference.

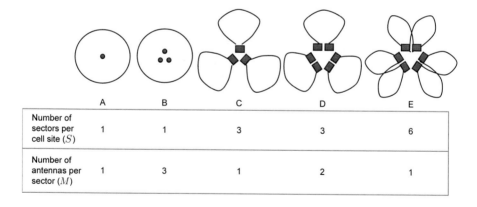

	A	B	C	D	E
Number of sectors per cell site (S)	1	1	3	3	6
Number of antennas per sector (M)	1	3	1	2	1

Fig. 1.12 Examples of different antenna configurations for a cell site.

ing the spectral efficiency. For example, slots can be allocated in a round-robin fashion among three users such that each user accesses the subcarrier on every third slot. Under CDMA with direct sequence spread spectrum, users access the channel with different code sequences. These sequences are designed to have low cross-correlation so that interference is minimized.

The latest cellular standards use orthogonal frequency division multiple-access (OFDMA). Here, as in FDMA, the bandwidth is partitioned into multiple subcarriers. However unlike FDMA, where a guard band is inserted between subcarriers to prevent inter-carrier interference, OFDMA improves the spectral efficiency by using an inverse Fast Fourier Transform (FFT) to multiplex signals across multiple subcarriers. As a result, signals overlap in the frequency domain yet are mutually orthogonal.

The 3GPP Long-term evolution (LTE) system [4] is an example of a current cellular standard that uses OFDMA. It can support channel bandwidths up to 20 MHz and partitions it into subcarriers of bandwidth 15 kHz. The time domain consists of 1 ms subframes spanning 14 symbol periods of duration 0.07 ms. The minimum unit of resource allocation is known as a *physical resource block* (PRB), which consists of 12 contiguous subcarriers (180 kHz bandwidth) in the frequency domain and 1 subframe in the time domain. The time and frequency resources for OFDMA can be represented as a two-dimensional grid as illustrated in Figure 1.13, where each segment corresponds to a PRB. Another OFDMA-based standard that uses a similar structure but with different parameters and terminology is IEEE 802.16e [5]. For simplicity we will use the LTE terminology in our discussion.

Scheduling and resource allocation

First- and second-generation cellular networks typically used circuit-switched communication to provide voice service over constant bit rate connections. In contrast, contemporary packet-based systems such as LTE and IEEE 802.16e employ packet switching where all traffic, regardless of content or type, is packaged as blocks of data known as packets. Each base (sector) employs a *scheduler* that allocates spectral resources for packet transmission among its assigned users. A separate scheduler is used for the uplink and downlink, and an independent allocation of resources can occur as often as once every subframe. As shown in Figure 1.13, the allocation of the resources is dynamic such that each PRB or group of PRBs can be allocated to a different user on each subframe. An encoding block for a given user occurs over a block

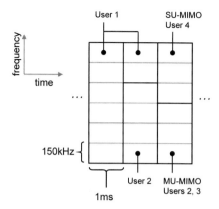

Fig. 1.13 Under the LTE standard, OFDMA is used, and the time/frequency resources are partitioned into physical resource blocks with bandwidth 180 kHz and 1 ms subframe time intervals. A scheduler allocates resources among users. MIMO techniques can be implemented for each resource block. For example, on the rightmost subframe, user 4 is allocated 3 resource blocks with each supporting SU-MIMO spatial multiplexing. Users 2 and 3 are spatially multiplexed on each of the lower resource blocks.

of symbols that spans one subframe in the time domain and one or more resource blocks in the frequency domain.

The scheduler allocates resources in order to maximize the total throughput while meeting quality of service (QoS) requirements for each user's application. These requirements include the average data rate and the maximum tolerable latency. For example, a large file transfer would require a large rate and could tolerate a large latency whereas a streaming audio application would require a lower rate and smaller latency.

For each transmission, the scheduler determines the appropriate transmission rate, which is based on measured channel characteristics and QoS requirements, and characterized by the coding rate and symbol modulation. This technique is known as *rate adaptation*. For downlink data transmission, a downlink control channel transmitted on resources orthogonal to the data traffic channel notifies the users of the modulation and coding scheme. For uplink data transmission, the downlink control channel notifies the scheduled users which resources to use and what the modulation and coding scheme should be.

In summary, the cellular infrastructure consists of cell sites that are partitioned into sectors. For a given sector, a base station communicates with multiple users and the sharing of spectral resources between the users is defined by the multiple-access technique. In contemporary networks, multiple-access is based on OFDMA, where the channel bandwidth is partitioned into narrowband subcarriers and time is partitioned into subframes. During each subframe, a base station scheduler allocates contiguous blocks of subcarriers (PRBs) to different users to optimize its performance metric.

If multiple antennas are available at the base and to mobile users, spatial multiplexing can be implemented over the PRBs. As shown in the rightmost subframe of Figure 1.13, user 4 is allocated 3 PRBs, and SU-MIMO spatial multiplexing is implemented for each PRB. Equation (1.1) can be used to model the received signal for either the uplink or downlink. Users 2 and 3 are spatially multiplexed on each of the lower 3 PRBs. If the transmission occurs for the uplink, equation (1.2) can be used to model the received signal; for the downlink, equation (1.3) can be used.

1.2.2 Co-channel interference

In the discussion of SU and MU-MIMO channel capacity in Section 1.1, we assumed that the links were corrupted by additive noise. The performance was characterized by the signal-to-noise ratio. In cellular systems, the link performance is dependent on the resource allocation between bases and is affected by *co-channel interference* arising from transmissions occuring on common frequency channels. If we treat the interference as an additional source of noise, the performance can be characterized by the *signal-to-interference-plus-noise ratio* (SINR).

As shown in Figure 1.14, interference could occur between cells and between sectors belonging to the same cell site. On the uplink, a base could receive *intercell* interference from co-channel users assigned to other bases, and on the downlink, a user could receive intercell interference from bases not assigned to it. *Intracell* interference could occur between sectors belonging to the same cell site as a result of power leaking through non-ideal sector side-lobes. Within a sector, interference could occur as a result of multiuser spatial multiplexing if the spatial channels are not orthogonal. Other than spatial interference, we assume there is no other intrasector interference within a given time-frequency resource block.

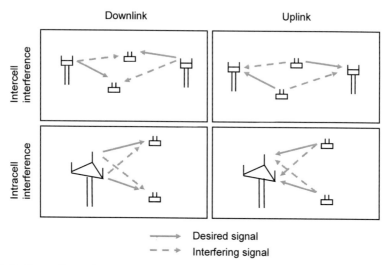

	Downlink	Uplink

Fig. 1.14 Intercell interference occurs between cells. Intracell interference occurs between sectors belonging to the same cell site. Intrasector interference occurs between spatially multiplexed users belonging to the same sector.

SINR and geometry

For the SU-MIMO channel in Section 1.1, we made a distinction between the channel SNR, defined as P/σ^2, and the SNR at the input of the decoder which is a function of P/σ^2, the channel realization $\mathbf{H}$, and the transceiver technique which defines the precoding and combining. For a cellular network, the SINR at the input of the decoder is a function of the transceivers and the transmit powers and channel realizations of the desired and interfering sources. Analogous to the channel SNR for the SU-MIMO channel, we define the *geometry* as the average received power from the desired source divided by the sum of the average received noise and interference power. The geometry accounts for the distance-based pathlosses and shadow fading realizations but is independent of the number of antennas. For example, on the downlink, a user at the edge of a cell is said to have low geometry if the average power received from its serving base is low compared to the co-channel interference power received from all other bases. On the other hand, a user very close to its assigned base has a high geometry if its received power from the desired base is high compared to the interference power.

As we will see, in typical cellular networks, the range of the geometry is about -10 dB to 20 dB for both the uplink and downlink. We can interpret the performance results in Figures 1.8 and 1.9 in a cellular context by relabeling the channel SNR on the x-axis with the geometry. For the SU-MIMO capacity performance, a geometry of 20 dB corresponds to a user positioned very close to its serving base. For the MU-MIMO sum-capacity performance, a geometry of 20 dB corresponds to all four users positioned very close to their serving base.

Impact of interference on capacity

Suppose that a user on the downlink receives average power P watts from its desired base and average total power αP $(0 \leq \alpha < 1)$ from the interfering bases. In the presence of additive Gaussian noise power σ^2, the geometry is $P/(\sigma^2 + \alpha P)$. Assuming there is no Rayleigh fading, the corresponding capacity (in bps/Hz) is

$$\log_2 \left(1 + \frac{P}{\sigma^2 + \alpha P} \right). \tag{1.26}$$

With α fixed, increasing the transmit power of both the desired and interfering bases by β watts results in a higher capacity because

$$\frac{P + \beta}{\sigma^2 + \alpha(P + \beta)} > \frac{P}{\sigma^2 + \alpha P} \tag{1.27}$$

for any $\beta > 0$. As P approaches infinity, the capacity approaches the limit $\log_2(1 + 1/\alpha)$ for $\alpha > 0$.

If the noise power is significantly greater than the interference power (i.e., $\alpha P \ll \sigma^2$), the performance is said to be *noise-limited*. In this case, increasing P results in a linear capacity gain if P/σ^2 is small and a logarithmic gain if P/σ^2 is large. On the other hand, if the interference power is significantly greater than the noise power (i.e., $\alpha P \gg \sigma^2$), the performance is *interference-limited*. As P increases, the capacity approaches its limit of $\log_2(1 + 1/\alpha)$.

Figure 1.15 shows an example of link capacity (1.26) versus SNR P/σ^2 for $\alpha = 0, 0.1, 0.5$. When there is no interference ($\alpha = 0$), the performance is noise limited for the entire range of SNR. For $\alpha = 0.1$, the performance is noise limited for the lower range of SNR and interference limited for SNRs greater than 20 dB. For $\alpha = 0.5$, the performance is interference-limited for

SNRs greater than 10 dB. It is desirable to operate a system so that the SNR is at the transition point between noise and interference-limited performance. At this point (for example, operating at an SNR of 10 dB for $\alpha = 0.5$), the power is used efficiently and increasing P further would not result in significant performance gains. If on the other hand P is fixed but interference can be mitigated to reduce α, significant performance gains can be achieved in the interference-limited SNR regime. For example in Figure 1.15, the gains from reducing α from 0.5 to 0.1 are significant at 20 dB SNR but insignificant at 0 dB SNR.

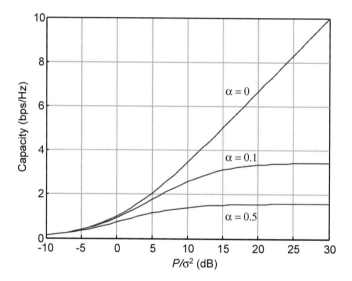

Fig. 1.15 Link capacity (1.26) versus SNR P/σ^2 with interference scaling factor α. If $\alpha = 0$, the performance is noise-limited, and the capacity increases without bound as SNR increases. If $\alpha > 0$, the performance is interference limited, and the capacity saturates as SNR increases.

Interference mitigation techniques

As discussed in Section 1.2.1, reduced frequency reuse can mitigate interference between cells by ensuring adjacent cells are assigned to orthogonal subcarriers. Other techniques for interference mitigation include the following.

- **Power control.** Transmit powers could be adjusted to meet target SINR requirements. Transmitters closer to their intended receiver could achieve their target with less power, resulting in less interference caused to other cells. Power control is often used in circuit-based systems to ensure the SINR is high enough to support reliable voice communication.
- **Interference averaging.** In CDMA standards, transmissions are spread over a wide bandwidth using pseudo-random code spreading or frequency hopping. The resulting interference power is averaged over the wider bandwidth, resulting in interference statistics with lower variance.
- **Soft handoff.** Interference can be reduced by allowing a terminal to communicate simultaneously with two or more bases over common spectral resources. On the downlink, adjacent bases send the same information to the user, and these signals can be combined to improve reliability. On the uplink, a similar diversity advantage can be achieved by detecting a user's signal independently at multiple bases and having a central controller decide which signal is the most reliable. Soft handoff is spectrally inefficient because the resources could have been used to serve additional users.

These three mitigation techniques can be implemented using a single antenna at the base and mobile. Using multiple antennas, interference can be mitigated by exploiting the *spatial domain*. The transmitter precoding techniques described in Section 1.1 create spatial channels that focus the beam towards the desired user while implicitly reducing the interference to any other receiver lying outside the spatial channel. If, on the downlink, a base has knowledge of the CSI of users assigned to other bases, it can intentionally steer transmissions away from those users to mitigate interference. Similarly, if on the uplink a base has knowledge of the CSI from interfering users, it can mitigate their interference through judicious receiver combining. In essence, multiuser detection is applied by a base jointly to assigned and interfering users. Because the MIMO techniques can create only a limited number of interference-free spatial channels, an important design decision for optimizing system performance is to determine the allocation of spatial resources between the assigned and interfering users. Assigning more resources to the assigned users potentially increases the multiplexing gain but the SINR of each user could be lower as a result.

The performance of interference mitigation techniques can be improved by coordinating the transmission and reception of base stations. Coordination strategies require enhanced baseband processing and additional backhaul network resources for sharing information between the bases, and the performance gains from coordination must be weighed against the cost of the added complexity. Spatial interference mitigation will be discussed in more detail in Chapters 5 and 6.

1.3 Overview of the book

A goal of this book is to apply theoretical Shannon capacity results to the practical performance of contemporary cellular networks. We do so by addressing the central problem of designing a MIMO cellular network to maximize the throughput spectral efficiency per unit area. Our method is to develop an understanding of MIMO-channel capacity metrics and to evaluate them in a simplified simulation environment that captures key features of real-world networks that are most relevant to the physical-layer performance. The simulation environment is not specific to a particular standard and is broad enough to capture characteristics of general contemporary packet-based networks.

During the course of this book, we hope to illuminate general principles for MIMO communication and wireless system design that can also be applied to networks beyond the cellular realm. Some of the questions relating to MIMO performance include the following. How does the capacity of a MIMO link scale with the number of antennas at both ends, in various SNR regimes? What transmitter and receiver architectures are required to achieve capacity? What is the impact of channel state information at the transmitter? In multiuser MIMO channels, how does the sum-capacity scale with the number of antennas or the number of users? How far from optimal are the simpler transmitter and receiver architectures that are more suitable for practical implementation?

Further, in the context of a cellular network, what is the best way to use multiple antennas, keeping the impact of intercell interference in mind? How can multiple antennas be used to mitigate intercell interference? When is single-user MIMO preferable to multiuser MIMO? How do MIMO techniques compare to simpler alternatives for improving area spectral efficiency such as sectorization?

The book is divided into two main parts. The first part consists of Chapters 2 through 4 and covers the fundamentals of communication over SU- and MU-

MIMO channels. The second part consists of Chapters 5 through 7 and applies the fundamental techniques in the design and operation of cellular networks. Figure 1.16 gives an overview of the chapter contents, channel models, and metrics.

- **Chapter 2** discusses the capacity of the SU-MIMO channel. We describe techniques for achieving capacity and also suboptimal techniques including linear receivers, space-time coding, and limited-feedback precoding.
- **Chapter 3** focuses on optimal techniques for MU-MIMO channels. The capacity regions and capacity-achieving techniques for the MAC and BC are described. Concepts useful for numerical performance evaluation are introduced, including a duality relating the BC and MAC capacity regions, and the numerical techniques for maximizing sum-rate performance metrics. Asymptotic sum-rate performance results provide insights into how the optimal MU-MIMO techniques should be implemented in cellular networks.
- **Chapter 4** describes suboptimal techniques for the MAC and BC. Due to the challenges of capacity-achieving techniques for the BC, most of the chapter focuses on various suboptimal BC precoding techniques.
- **Chapter 5** shifts the discussion from isolated MIMO channels to cellular networks, and develops a general simulation methodology for evaluating cellular network performance that is applicable to any contemporary packet-switched cellular standard. This chapter also describes practical aspects of these networks including scheduling and procedures for acquiring channel state information.
- **Chapter 6** presents numerical simulation results for a variety of network architectures. It consists of three sections: single-antenna bases, multiple-antenna bases, and coordinated bases. This chapter synthesizes material from the previous chapters and concludes with a flowchart that gives design recommendations for applying MIMO in cellular networks.
- **Chapter 7** describes specific multiple-antenna techniques for existing and future cellular standards. We consider how the general principles and techniques discussed in the other chapters are specifically implemented in these standards.

Channel type	Chapter	Channel model	Metrics
Single-user MIMO	2	(M,N) MIMO channel $x = \mathbf{H}s + \mathbf{n}$ $N \times M$	**Capacity** $C^{\mathrm{OL}}(\mathbf{H},P), C^{\mathrm{CL}}(\mathbf{H},P)$
Multiuser MIMO	3, 4	$((K,N),M)$ multiaccess channel (MAC) $x = \sum_{k=1}^{K} \mathbf{H}_k s_k + \mathbf{n}$ $M \times N$ $(M,(K,N))$ broadcast channel (BC) $x_k = \sum_{k=1}^{K} \mathbf{H}_k^H s + \mathbf{n}_k$ $N \times M$	**Capacity regions** $C^{\mathrm{MAC}}(\{\mathbf{H}_k\},\{P_k/\sigma^2\})$ $C^{\mathrm{BC}}(\{\mathbf{H}_k^H\},P/\sigma^2)$ **Sum-rate capacities** $C^{\mathrm{MAC}}(\{\mathbf{H}_k\},\{P_k/\sigma^2\})$ $C^{\mathrm{BC}}(\{\mathbf{H}_k^H\},P/\sigma^2)$
Cellular network	5, 6	Uplink channel, base b received signal $x_b = \sum_{j \in U_b} \mathbf{H}_j s_j + \sum_{j \notin U_b} \mathbf{H}_j s_j + \mathbf{n}_b$ $M \times N$ Downlink channel, user k received signal $x_k = \mathbf{H}_{k,b'}^H s_{b'} + \sum_{b \neq b'} \mathbf{H}_{k,b}^H s_b + \mathbf{n}_k$ $N \times M$	**Proportional-fair criteria** Mean throughput per site Peak user rate Cell-edge user rate **Equal-rate criteria** Throughput per site

Fig. 1.16 Overview of chapter contents, channel models, and metrics.

Chapter 2
Single-user MIMO

In this chapter we study the single-user MIMO channel which is used to model the communication link between a base station and a user. We explore in more detail the fundamental results that were briefly described in the previous chapter. We derive the open-loop and closed-loop MIMO channel capacities and describe techniques for achieving capacity including architectures known as V-BLAST and D-BLAST. We also describe classes of suboptimal techniques such as linear receivers, space-time coding for transmitting a single data stream from multiple antennas, and precoding when there is limited knowledge of CSI at the transmitter.

2.1 Channel model

Figure 2.1 shows the baseband model for a single-user (M, N) MIMO link with M transmit and N receive antennas. A stream of data bits is communicated over the channel. We let $d^{(i)} \in \{+1, -1\}$ represent the data bit with index $i = 0, 1, \ldots$. The data stream is processed to create a sequence of transmitted data symbols. We let $s_m^{(t)} \in \mathbb{C}$ denote the complex baseband signal transmitted from antenna m during period t. For a given symbol period t, the channel between the mth $(m = 1, \ldots, M)$ transmit antenna and the jth $(j = 1, \ldots, N)$ receive antenna is characterized by a scalar value $h_{j,m}^{(t)} \in \mathbb{C}$ which represents the complex amplitude of the narrowband, frequency-nonselective channel. Because each receive antenna is exposed to all transmit antennas, the baseband signal received at antenna j during time t can be written as a linear combination of the transmitted signals:

$$x_j^{(t)} = \sum_{m=1}^{M} h_{j,m}^{(t)} s_m^{(t)} + n_j^{(t)}, \tag{2.1}$$

where $n_j^{(t)}$ is complex additive noise. By stacking the received signals from all N antennas in a tall vector, we can write:

$$\begin{bmatrix} x_1^{(t)} \\ \vdots \\ x_N^{(t)} \end{bmatrix} = \begin{bmatrix} h_{1,1}^{(t)} & \cdots & h_{1,M}^{(t)} \\ \vdots & \ddots & \vdots \\ h_{N,1}^{(t)} & \cdots & h_{N,M}^{(t)} \end{bmatrix} \begin{bmatrix} s_1^{(t)} \\ \vdots \\ s_M^{(t)} \end{bmatrix} + \begin{bmatrix} n_1^{(t)} \\ \vdots \\ n_N^{(t)} \end{bmatrix} \tag{2.2}$$

which can be written in the more compact form

$$\mathbf{x}^{(t)} = \mathbf{H}^{(t)} \mathbf{s}^{(t)} + \mathbf{n}^{(t)}, \tag{2.3}$$

where $\mathbf{x}^{(t)} \in \mathbb{C}^{N \times 1}$, $\mathbf{H}^{(t)} \in \mathbb{C}^{N \times M}$, $\mathbf{s}^{(t)} \in \mathbb{C}^{M \times 1}$, and $\mathbf{n}^{(t)} \in \mathbb{C}^{N \times 1}$. For simplicity, we will typically drop the time index t. The noise vector $\mathbf{n}$ is assumed to be zero-mean, spatially white (ZMSW), circularly symmetric, additive complex Gaussian, with each component having variance σ^2: $\mathbb{E}(\mathbf{n} \mathbf{n}^H) = \sigma^2 \mathbf{I}_N$, where $\mathbf{I}_N$ is the $N \times N$ identity matrix. (When the noise is spatially colored, i.e., when the covariance of $\mathbf{n}$ is not a multiple of the identity matrix, we can suppose that the receiver whitens the noise first, by multiplying the received signal vector by the inverse square root of the noise covariance.) The components of the signal vectors $\mathbf{s}^{(t)}$, $t = 1, 2, \ldots$ are the encoded symbols obtained by processing an information bit stream $d^{(1)}, d^{(2)}, \ldots$ which we denote by $\{d^{(i)}\}$. The signal vector is modeled as a stationary random process with zero mean $\mathbb{E}(\mathbf{s}) = \mathbf{0}_M$ and covariance $\mathbf{Q} := \mathbb{E}(\mathbf{ss})^H$. The signal is subject to the power constraint $\operatorname{tr} \mathbf{Q} = \mathbb{E}(\|\mathbf{s}\|^2) = P$.

The realization $\mathbf{H}^{(t)}$ is drawn from a stationary, ergodic random process to model the fading of the wireless channel. Due to the movement of the transmitter, receiver, and local scatterers, the signal transmitted from antenna m and received by antenna j experiences multipath fading caused by varying path lengths to the scatterers. As a result of the central limit theorem, the complex amplitude of the combined multipath signals can be modeled as a complex Gaussian random variable. If the spacing between the M transmit antennas is sufficiently large relative to the channel angle spread (which is determined by the height of the antennas relative to the height of the local scatterers), then the M channel coefficients $h_{j,1}^{(t)}, \ldots, h_{j,M}^{(t)}$ for receive antenna j will be uncorrelated. Likewise, if the spacing between the N receive antennas is sufficiently large, then the N channel coefficients $h_{1,m}^{(t)}, \ldots, h_{N,m}^{(t)}$ for transmit antenna m will be uncorrelated. A channel in which the coefficients of $\mathbf{H}^{(t)}$ are uncorrelated (or weakly correlated) is said to be *spatially rich*.

Typically, we will assume in this book that for a given symbol interval t, the elements of $\mathbf{H}^{(t)}$ are not only spatially rich but also independent and identically distributed (i.i.d.) complex Gaussian random variables with zero mean and unit variance. Because the amplitude of each element has a Rayleigh distribution, this channel distribution is known as an *i.i.d. Rayleigh* distribution. As a result of the channel normalization, the average received signal power is P, and the signal-to-noise ratio (SNR), defined as the ratio of the received signal power and noise power, is P/σ^2.

With regard to the time evolution of the channel realizations, we define two types of channel model.

1. **Fast-fading**: the channel changes fast enough between symbol periods that each coding block effectively spans the entire distribution of the random process (i.e., ergodicity holds).
2. **Block-fading**: the channel is fixed for the duration of a coding block, but it changes from one block to another.

In practice, coding block lengths are on the order of a millisecond, so users with low mobility (stationary or pedestrian users) experience slowly fading channels consistent with the block-fading model. In this book, we focus mainly on the block-fading model. Further, we usually assume an i.i.d. Rayleigh fading model for the channel.

In the rest of this section, we briefly describe more general channel models that account for propagation environments that are not spatially rich and therefore induce correlated fading across transmitter and receiver antenna pairs.

2.1.1 Analytical channel models

Analytical channel models attempt to describe the end-to-end transfer functions between the transmitting and receiving antenna arrays by accounting for physical propagation and antenna array characteristics [6]. Most analytical channel models capture the various propagation mechanisms through the correlations of the random channel coefficients. Below we describe the most well-known correlation-based analytical MIMO channel models.

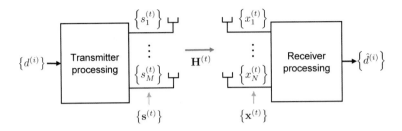

Fig. 2.1 The (M, N) single-user MIMO link. A stream of data bits $\{d^{(i)}\}$ is processed to form a stream of encoded symbol vectors $\{\mathbf{s}^{(t)}\}$. The signal is transmitted from M antennas over the channel $\mathbf{H}$ and is received by N antennas. The received signal $\{\mathbf{x}^{(t)}\}$ is processed to provide estimates of the data stream bits $\{\hat{d}^{(i)}\}$.

2.1.1.1 Kronecker MIMO channel model

The Kronecker MIMO channel model is probably the best-known correlation-based model and stems from early efforts in the community [7–11] to find models which correspond to a given pair of transmission and receiver correlation matrices (denoted for simplicity as $\mathbf{R}_T = \mathbb{E}\left(\mathbf{H}^H \mathbf{H}\right)$ and $\mathbf{R}_R = \mathbb{E}\left(\mathbf{H}\mathbf{H}^H\right)$, respectively). It hinges on the assumption that these two correlation matrices are *separable*. Mathematically, this assumption is expressed as

$$\mathbf{R_H} = \mathbf{R}_T \otimes \mathbf{R}_R, \qquad (2.4)$$

where $\mathbf{R_H}$ is defined as $\mathbf{R_H} = \mathbb{E}\left(\text{vec}\left(\mathbf{H}\right)\text{vec}\left(\mathbf{H}\right)^H\right)$, where the vec operator stacks the columns of the operand matrix vertically, and $\otimes$ denotes the Kronecker product between two matrices. It can be shown that, in this case, the channel matrix can be expressed as

$$\mathbf{H} = \mathbf{R}_R^{1/2}\mathbf{H}_{\text{i.i.d.}}\mathbf{R}_T^{1/2} \qquad (2.5)$$

where $\mathbf{H}_{\text{i.i.d.}}$ is a $N \times M$ matrix of i.i.d. circularly symmetric complex Gaussian random variables of zero mean and unit variance. The assumption of separable transmit/receive correlations of course limits the generality of this model, as it is unable to capture any coupling between direction of departure and direction of arrival spectra. Examples of simple channel models that are not captured by the Kronecker channel model are the so-called "keyhole channel," as well as the single- and double-bounce models [12] described below.

However, the Kronecker model has been very popular due to its successful role in quantifying MIMO capacity of correlated channels as well as to its modularity, which allows separate transmitter and receiver array optimization.

2.1.1.2 Single-bounce analytical MIMO channel model

In this case we assume that the signal transmitted from each transmitter bounces once off each of a set of (say V) scatterers before it reaches any receiver antenna. In this single-bounce case, the MIMO channel can be modeled as

$$\mathbf{H}_{\mathrm{SB}}\left(N, M, V\right) = \boldsymbol{\Phi}_{\mathrm{R}}\left(N, V\right) \mathbf{H}_{\mathrm{i.i.d.}}\left(V, V\right) \boldsymbol{\Phi}_{\mathrm{T}}^{H}\left(M, V\right), \qquad (2.6)$$

where $\boldsymbol{\Phi}_R$ and $\boldsymbol{\Phi}_T$ are matrices that define the electrical path lengths from the V scatterers and the N, M antenna elements, respectively. In fact, it turns out that these matrices are the matrix square roots of the correlation matrices on each side of transmission, i.e.

$$\mathbf{H}_{\mathrm{SB}}\left(N, M, V\right) = \mathbf{R}_{\mathrm{R}}^{1/2}\left(N, N\right) \mathbf{H}_{\mathrm{i.i.d.}}\left(N, M\right) \mathbf{R}_{\mathrm{T}}^{1/2}\left(M, M\right). \qquad (2.7)$$

In the above, the first two arguments within the parentheses denote the matrix dimensions; the third argument, when present, denotes the assumed number of scatterers. The model in (2.7) has been used successfully to characterize many practical cases where correlation among antenna elements on each side of the link is present (e.g. due to their proximity or a limited angle spread), despite the fundamental richness of the in-between propagation environment. In other words, it models local correlation well. It should be noted of course that when V is smaller than $\min\left(M, N\right)$, the channel in (2.7) will suffer severe degradation in its richness, as it will lose rank (notice that such a phenomenon cannot be captured by the Kronecker model in (2.5)). To maximize richness, V should be greater than or equal to NM. The middle ground between $\min\left(M, N\right)$ and NM provides intermediate levels of richness (see [12] for some simulated results). It should also be noted that smaller scale effects, such as those due to mutual coupling, are not captured in this model (see [13, 14]).

2.1.1.3 Double-bounce and keyhole MIMO analytical channel models

In some cases, channel richness is compromised by the fact that some waves follow common paths, thus limiting the independence between some signals; as noted in [8, 15], this could be due either to a separation in free space or to some sort of wave-guiding effect. A model that captures these effects is the so-called double-bounce model, which is an extension of the single-bounce model that includes a second ring of scatterers and is described by the following equation:

$$\mathbf{H}_{\mathrm{DB}}\left(N, M, V\right) = \mathbf{\Phi}_{\mathrm{R}}(N, V_1)\mathbf{H}_{\mathrm{i.i.d.}}\left(V_1, V_1\right)\mathbf{X}\left(V_1, V_2\right)\mathbf{\Phi}_{\mathrm{T}}(M, V_2)^H. \quad (2.8)$$

where V_1 and V_2 denote the number of scatterers in the first and second ring, respectively. A special case of the model in (2.8), where the resulting matrix has only a single nonzero eigenvalue, is the so-called *keyhole* or *pinhole* channel [8, 15].

2.1.1.4 The Weichselberger MIMO analytical channel model

This model attempts to relax the separability between transmitter and receiver correlations by exploiting the eigenvalue decomposition of the corresponding correlation matrices, shown below:

$$\begin{aligned}\mathbf{R}_T &= \mathbf{U}_T\mathbf{\Lambda}_T\mathbf{U}_T^H \\ \mathbf{R}_R &= \mathbf{U}_R\mathbf{\Lambda}_R\mathbf{U}_R^H\end{aligned}, \quad (2.9)$$

where $\mathbf{U}_T$, $\mathbf{U}_R$ are unitary and $\mathbf{\Lambda}_T$, $\mathbf{\Lambda}_R$ are diagonal matrices. The Weichselberger MIMO channel model is given by the following expression:

$$\mathbf{H} = \mathbf{U}_R\left(\mathbf{\Omega} \bullet \mathbf{H}_{\mathrm{i.i.d.}}\right)\mathbf{U}_T^H, \quad (2.10)$$

where $\mathbf{\Omega}$ is a $N \times M$ coupling matrix that determines the average power coupling between the transmit and receive eigenmodes and $\bullet$ denotes the Schur-Hadamard product (element-wise multiplication). In fact, the Kronecker model is a special case of the Weichselberger model where the coupling matrix $\mathbf{\Omega}$ has rank 1. Other classes of random analytical MIMO channel models include propagation-based versions, such as:

- The finite scatterer model, which assumes a finite number of scatterers and models the angles of departure and arrival, scattering coefficient and delay for each scatterer [16]. This model allows incorporation of both single-bounce and double-bounce scattering.
- The maximum entropy model [17], which attempts to incorporate properties of the propagation environment and system parameters via the maximum entropy principle so as to maximize the model's match to the a priori known information about the link.
- The virtual channel model which exploits the so-called "deconstructed" MIMO channel representation proposed in [18], capturing the "inner" propagation environment between virtual transmission and reception scatterers.

2.1.1.5 The Ricean MIMO channel model

Similarly to the case of scalar channels, when a line of sight (LOS) exists between the transmitter and receiver, the channel is modeled as the sum of a random part representing the non-LOS component and a deterministic part that represents the LOS component. The well-known scalar Ricean channel can be extended to the MIMO case as follows:

$$\mathbf{H} = \frac{\mathbf{H}_{\mathrm{R}} + \sqrt{K}\mathbf{H}_{\mathrm{D}}}{\sqrt{1+K}}, \tag{2.11}$$

where $K \geq 0$ is the Rice factor (also called the "K factor"), $\mathbf{H}_{\mathrm{D}}$ denotes the LOS deterministic channel matrix and $\mathbf{H}_{\mathrm{R}}$ denotes the random channel matrix that can be modeled according to any of the MIMO channel models presented above.

2.1.2 Physical channel models

In contrast to the analytical channel models, physical channel models focus on the properties of the physical environment between transmitter and receiver array. Two classes of physical channel model are briefly described below:

- **Ray-tracing models**
  Ray-tracing (RT) models (see [6, 19]) are widely considered to be among the most reliable deterministic channel models for wireless communica-

tions: they rely on the theory of geometrical optics to predict the way the electromagnetic waves will reach the receiver after they interact with the environment's obstacles (which cause reflection, absorption, diffraction, and so on). The Achilles' heal of the RT approach is the need to know in advance the physical obstacles of the propagation environment between the transmitter and receiver. MIMO extensions of the RT approach have been proposed in [20, 21]. In the MIMO case, the pattern and polarization of each antenna must be taken into account; however, this can be done in a modular fashion, making the technique applicable to any known antenna array configuration.

- **Geometry-based stochastic physical models**

  Contrary to the deterministic nature of the RT approach described above, which exploits the propagation environment's geometry in a deterministic fashion, geometry-based stochastic physical models (GSCM) model the scatterer locations in stochastic (random) terms, i.e. via their statistical distributions. Beyond Lee's original model of deterministic scatterer locations on a circle around the mobile [22], various random scatterer distributions (including scatterer clustering) have since been proposed in, for example, [23–26]. In the single-bounce approach each transmit/receive path is broken into two sub-paths: transmitter-to-scatterer and scatterer-to-receiver (described by their direction of departure, direction of arrival, and path distance); the scatterer itself is modeled typically via the introduction of a random phase shift. Multiple-bounce scattering has also been proposed in order to address more complicated propagation environments (see [27–29]). As mentioned above, MIMO versions of these models are derived by considering the specific configuration and characteristics of the antenna arrays on each side of the link.

2.1.3 Other extensions

The models mentioned above typically assume narrow-band propagation; in other words, they are frequency-flat. Several wideband extensions have been proposed in the literature to capture broadband communication links (see e.g. references in [6]); these are especially relevant in view of the emerging LTE/LTE Advanced and WiFi/WiMAX type systems, which typically use OFDM and operate in bandwidths on the order of several tens of MHz. Also, it was assumed that the propagation channel is time-invariant; time-varying extensions have also been proposed [30, 31]; these are especially relevant in cases of high user mobility. Mutual coupling between antenna elements (in particular its role in affecting the channel's spatial correlation properties) in the above representative classes of MIMO channel models have been proposed in [13, 14]. Finally, it should be mentioned that large collaborative efforts have been undertaken over the last decade or so in order to propose MIMO channel models that are fit for current and emerging wireless standards. These include the COST259 [32], COST273 [29], IEEE 802.11n [33], Hiperlan2 [34], Stanford University Interim (SUI) [26] and IEEE 802.16 [35]. The Spatial Channel Model described in [36] is used as the basis for standards-related simulations for 3GPP and 3GPP2.

2.2 Single-user MIMO capacity

In this section, expanding on the brief discussion on SU-MIMO capacity in Chapter 1, we derive the capacity for the single-user MIMO channel. We first derive the open-loop and closed-loop MIMO capacity for a fixed channel realization, and then we study the performance of the capacity averaged over random channel realizations.

2.2.1 Capacity for fixed channels

To begin with, we will focus on the case where the channel matrix $\mathbf{H}^{(t)}$ in (2.3) equals some fixed matrix $\mathbf{H} \in \mathbb{C}^{N \times M}$ for all t, i.e., the channel is time-invariant. We will further assume that $\mathbf{H}$ is known exactly to both the transmitter and receiver. After dropping the time index t in (2.3) for convenience, the input-output relationship of the channel reduces to

$$\mathbf{x} = \mathbf{Hs} + \mathbf{n}. \qquad (2.12)$$

The channel input $\mathbf{s}$ is subject to an average power constraint of P. The additive noise vector $\mathbf{n}$ has a circularly symmetric complex Gaussian distribution with zero mean and covariance $\sigma^2 \mathbf{I}_N$.

By Shannon's channel coding theorem [37], the capacity $C\left(\mathbf{H}, P/\sigma^2\right)$ of the above channel, defined as the maximum data rate at which the decoding error probability at the receiver can be made arbitrarily small with sufficiently long codewords, is given by the maximum mutual information $I(\mathbf{s}; \mathbf{x})$ between the input $\mathbf{s}$ and the output $\mathbf{x}$, over all possible distributions for $\mathbf{s}$ that satisfy the power constraint $\operatorname{tr}\left(E\left[\mathbf{ss}^H\right]\right) \leq P$. There is no loss of optimality here in restricting $\mathbf{s}$ to have zero mean, since $\mathbf{s} - E\left[\mathbf{s}\right]$ automatically satisfies the power constraint if $\mathbf{s}$ does, and also yields the same mutual information with $\mathbf{x}$ as $\mathbf{s}$. Now,

$$I(\mathbf{s}; \mathbf{x}) = h(\mathbf{x}) - h\left(\mathbf{x} \mid \mathbf{s}\right) \tag{2.13}$$

$$= h(\mathbf{x}) - h(\mathbf{n}), \tag{2.14}$$

where $h(\mathbf{z})$ denotes the differential entropy of the random vector $\mathbf{z}$, and $h\left(\mathbf{z} \mid \mathbf{y}\right)$ the conditional differential entropy of $\mathbf{z}$ given $\mathbf{y}$.

The differential entropies can be evaluated using the following important result about differential entropy (see, e.g., [38] for a proof): *If $\mathbf{z}$ is any zero-mean complex random vector with covariance $E\left[\mathbf{zz}^H\right] = \mathbf{R}_z$, then $h(\mathbf{z}) \leq \log\left|\pi e \mathbf{R}_z\right|$, with equality holding if and only if $\mathbf{z}$ has a circularly symmetric complex Gaussian distribution.*

Thus in (2.14), we have $h(\mathbf{n}) = \log\left|\pi e \sigma^2 \mathbf{I}_N\right|$. Maximizing $I(\mathbf{s}; \mathbf{x})$ therefore amounts to maximizing $h(\mathbf{x})$. Further, for any zero-mean $\mathbf{s}$ with covariance $E\left[\mathbf{ss}^H\right] = \mathbf{R}_s$, the channel output $\mathbf{x}$ is also zero-mean and has the covariance $\sigma^2 \mathbf{I}_N + \mathbf{H}\mathbf{R}_s\mathbf{H}^H$. Consequently,

$$h(\mathbf{x}) \leq \log\left|\pi e \left(\sigma^2 \mathbf{I}_N + \mathbf{H}\mathbf{R}_s\mathbf{H}^H\right)\right|, \tag{2.15}$$

with equality if and only if $\mathbf{x}$ is circularly symmetric complex Gaussian. The latter condition holds when the input $\mathbf{s}$ is itself circularly symmetric complex Gaussian. We can therefore conclude that, among all zero-mean input distributions with a given covariance $\mathbf{R}_s$, the one that maximizes $I(\mathbf{s}; \mathbf{x})$ is circularly symmetric complex Gaussian. Further, the corresponding mutual information is

$$I(\mathbf{s}; \mathbf{x}) = \log\left|\pi e \left(\sigma^2 \mathbf{I}_N + \mathbf{H}\mathbf{R}_s\mathbf{H}^H\right)\right| - \log\left|\pi e \sigma^2 \mathbf{I}_N\right| \tag{2.16}$$

$$= \log \frac{\left|\sigma^2 \mathbf{I}_N + \mathbf{H}\mathbf{R}_s\mathbf{H}^H\right|}{\left|\sigma^2 \mathbf{I}_N\right|} \tag{2.17}$$

$$= \log\left|\mathbf{I}_N + \left(1/\sigma^2\right) \mathbf{H}\mathbf{R}_s\mathbf{H}^H\right|. \tag{2.18}$$

Therefore the problem of determining the capacity $C\left(\mathbf{H}, P/\sigma^2\right)$ of the channel in (2.12) is reduced to that of finding the input covariance $\mathbf{R}_s$ that maximizes the RHS of (2.18), subject to the constraint $\mathrm{tr}\left(\mathbf{R}_s\right) \leq P$:

$$C\left(\mathbf{H}, P/\sigma^2\right) = \max_{\substack{\mathbf{R}_s \succeq 0 \\ \mathrm{tr}(\mathbf{R}_s) \leq P}} \log \left| \mathbf{I}_N + \left(1/\sigma^2\right) \mathbf{H}\mathbf{R}_s\mathbf{H}^H \right|. \qquad (2.19)$$

Clearly, in (2.19), any specific choice of the input covariance $\mathbf{R}_s$ satisfying $\mathrm{tr}\left(\mathbf{R}_s\right) \leq P$ will yield an achievable data rate, i.e., a lower bound on the channel capacity. One such choice that is often of interest is $\mathbf{R}_s = (P/M)\mathbf{I}_M$, which corresponds to an *isotropic* input, i.e., sending independent data streams at the same power from each of the transmit antennas. The corresponding achievable rate, which we will loosely term the "open-loop capacity" of the channel and denote by $C^{\mathrm{OL}}\left(\mathbf{H}, P/\sigma^2\right)$, is given by

$$C^{\mathrm{OL}}\left(\mathbf{H}, P/\sigma^2\right) = \log \left| \mathbf{I}_N + \frac{P}{M\sigma^2} \mathbf{H}\mathbf{H}^H \right|. \qquad (2.20)$$

In order to motivate the choice of an isotropic input and the concept of open-loop capacity, one can consider a situation where the transmitter has no knowledge of the channel matrix $\mathbf{H}$ (but the receiver still knows it perfectly). The isotropy of the additive noise $\mathbf{n}$ then suggests that the transmitter should employ an isotropic input, hedging against its ignorance of the channel by signaling with equal power in M orthogonal directions. More rigorous justifications can be given for the optimality of an isotropic input in the context of an ergodic channel model with spatially white noise [38].

2.2.1.1 Optimal input covariance

We will now sketch the derivation of the optimal input covariance $\mathbf{R}_s$ in (2.19). The key idea here is to show that the MIMO channel can be decomposed into several single-input single-output (SISO) channels that operate in parallel without interfering with each other, and must share the total available transmit power of P. The optimal power allocation between these SISO channels can then be obtained by a procedure commonly referred to as "waterfilling" (the reason for the name will soon become clear). While the derivation is of secondary importance for this book, we will provide this rough proof because it reveals the important concept of *spatial modes*.

The decomposition of the MIMO channel into non-interfering SISO channels is based on the *singular value decomposition* (SVD) of the $N \times M$ channel matrix $\mathbf{H}$. This decomposition allows us to express $\mathbf{H}$ as

$$\mathbf{H} = \mathbf{U}\mathbf{\Sigma}\mathbf{V}^H, \tag{2.21}$$

where $\mathbf{U}$ and $\mathbf{V}$ are $N \times N$ and $M \times M$ unitary matrices, respectively (so $\mathbf{U}\mathbf{U}^H = \mathbf{U}^H\mathbf{U} = \mathbf{I}_N$, $\mathbf{V}\mathbf{V}^H = \mathbf{V}^H\mathbf{V} = \mathbf{I}_M$) and $\mathbf{\Sigma}$ is an $N \times M$ diagonal matrix. Each element of $\mathrm{diag}(\mathbf{\Sigma})$ is a *singular value* of $\mathbf{H}$, i.e., the positive square root of an eigenvalue of either $\mathbf{H}\mathbf{H}^H$ (if $N \le M$) or $\mathbf{H}^H\mathbf{H}$ (if $N \ge M$). Moreover, the columns of $\mathbf{U}$ are eigenvectors of $\mathbf{H}\mathbf{H}^H$, and the columns of $\mathbf{V}$ are eigenvectors of $\mathbf{H}^H\mathbf{H}$.

Using (2.21) in (2.12), we get

$$\mathbf{x} = \left(\mathbf{U}\mathbf{\Sigma}\mathbf{V}^H\right)\mathbf{s} + \mathbf{n} \Rightarrow \mathbf{U}^H\mathbf{x} = \left(\mathbf{U}^H\mathbf{U}\right)\mathbf{\Sigma}\left(\mathbf{V}^H\mathbf{s}\right) + \mathbf{U}^H\mathbf{n} \Rightarrow \mathbf{x}' = \mathbf{\Sigma}\mathbf{s}' + \mathbf{n}', \tag{2.22}$$

where $\mathbf{x}' = \mathbf{U}^H\mathbf{x}$, $\mathbf{s}' = \mathbf{V}^H\mathbf{s}$, and $\mathbf{n}' = \mathbf{U}^H\mathbf{n}$. Note that $\mathbf{n}'$ has the same distribution as $\mathbf{n}$, since it is obtained by a unitary linear transformation of a zero-mean circularly symmetric complex Gaussian vector whose covariance is a multiple of the identity matrix. So the components of $\mathbf{n}'$ are all independent, circularly symmetric, complex Gaussian random variables of mean 0 and variance σ^2. Note also that the signal terms of (2.22) are *uncoupled*, due to the diagonal structure of $\mathbf{\Sigma}$.

Let us assume now that $\mathrm{rank}(\mathbf{H}) = r$ (where $r \le \min(M, N)$). The matrix $\mathbf{\Sigma}$ will then have r positive diagonal elements, which we will denote by $\lambda_i, i = 1, \ldots, r$. These are the singular values of $\mathbf{H}$, and $\lambda_i^2, i = 1, \ldots, r$ are the eigenvalues of $\mathbf{H}\mathbf{H}^H$. We will assume further that $\lambda_1 \ge \lambda_2 \ge \cdots \ge \lambda_r$. So (2.22) can equivalently be written as:

$$x_i' = \lambda_i s_i' + n_i', \quad i = 1, \ldots, r. \tag{2.23}$$

(If $r < N$ there are also $N - r$ equations of the type $x_i' = n_i'$, $i = r+1, \ldots, N$, which contain no input signal information, and can therefore be neglected.) Note that (2.23) describes an ensemble of r parallel, non-interfering SISO channels, with gains $\lambda_1, \lambda_2, \ldots, \lambda_r$ and noise variance σ^2. As a result, we can depict the equivalent signal model as shown in Figure 2.2.

Assuming now that the transmitter allocates power $P_i = E|s_i'|^2$ to the ith channel in (2.23), the SNR on the ith SISO channel is $\rho_i = \lambda_i^2 P_i/\sigma^2$, and the rate achievable over it is $R_i = \log_2(1 + \rho_i)$. The overall rate achieved

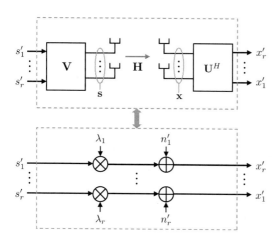

Fig. 2.2 Decomposition of the MIMO channel into r constituent SISO channels, where r is the rank of $\mathbf{H}$.

is $\sum_i R_i$, i.e., the sum of the rates over the individual SISO channels. The capacity of the MIMO channel is then obtained by maximizing the overall rate over all power allocations, subject to the total transmit power constraint:

$$C^{\mathrm{CL}}(\mathbf{H}, P/\sigma^2) = \max_{\substack{P_1, P_2, \ldots, P_r \\ \sum_i P_i = P}} \sum_{i=1}^{r} \log_2 \left(1 + \frac{P_i \lambda_i^2}{\sigma^2} \right). \qquad (2.24)$$

The objective function in (2.24) is concave in the variables P_i and can be maximized using Lagrangian methods (see [38]), yielding the following solution:

$$P_i^{\mathrm{Opt}} = \left(\mu - \frac{\sigma^2}{\lambda_i^2} \right)^+, \text{with } \mu \text{ chosen such that } \sum_{i=1}^{r} P_i^{\mathrm{Opt}} = P. \qquad (2.25)$$

Here $(a)^+ = \max(a, 0)$. The capacity of the channel is then given by [38]

$$C^{\mathrm{CL}}(\mathbf{H}, P/\sigma^2) = \sum_{i=1}^{r} \left[\log_2 \left(\frac{\lambda_i^2 \mu}{\sigma^2} \right) \right]^+. \qquad (2.26)$$

The covariance matrix of the transmitted signal $\mathbf{s}$ is given by:

$$\mathbf{R_s} = \mathbf{V} \left[\operatorname{diag}(P_1, \cdots, P_M) \right] \mathbf{V}^H. \qquad (2.27)$$

The optimal power allocation between the eigenmodes of the channel, given in (2.25), can be computed by a procedure known as "waterfilling" [37]. Water is poured in a two-dimensional container whose base consists of r steps, where the "height" of each step is σ^2/λ_j^2. If we let μ in (2.25) be the water level, the optimal power allocated to the jth eigenmode P_j^{Opt} is the difference between the water level and the height of the jth step (provided the water level is higher than that step), where μ is set so the total allocated power is P.

Figure 2.3 illustrates the waterfilling algorithm for a MIMO channel with three eigenmodes: $\lambda_1 > \lambda_2 > \lambda_3$. If SNR is very low, the water level, as indicated by the horizontal dotted line, covers only the first step. This indicates that all the power P is allocated to the dominant eigenmode ($P_1^{\text{Opt}} = P$) and no power is allocated to the others ($P_2^{\text{Opt}} = P_3^{\text{Opt}} = 0$). As the SNR increases, the other eigenmodes will be activated. For very high SNR, all three modes are activated, and the difference in height between the steps is insignificant compared to the water level. In this case, the power is allocated approximately equally among the three eigenmodes: $P_1^{\text{Opt}} \approx P_2^{\text{Opt}} \approx P_3^{\text{Opt}}$.

2.2.2 Performance gains

Having established the capacity of open- and closed-loop MIMO channels, we now discuss the performance gains of MIMO relative to conventional single-antenna techniques. In this section, we study in more detail the performance gains mentioned briefly in Chapter 1, namely that the MIMO gains in the low-SNR regime come about through antenna combining, and that the gains in the high-SNR regime come from spatial multiplexing. We also consider the capacity gains as the number of transmit and receive antennas increases without bound.

Under a block-fading channel model, the channel realization is random from block to block, and the capacity for each realization is a random variable. A useful performance measure is the *average capacity* obtained by taking the expectation of the capacity with respect to the distribution of $\mathbf{H}$. The average open-loop and closed-loop capacities are defined respectively as

$$\bar{C}^{\text{OL}}(M, N, P/\sigma^2) := \mathbb{E}_{\mathbf{H}} C^{\text{OL}}(\mathbf{H}, P/\sigma^2) \tag{2.28}$$

$$\bar{C}^{\text{CL}}(M, N, P/\sigma^2) := \mathbb{E}_{\mathbf{H}} C^{\text{CL}}(\mathbf{H}, P/\sigma^2). \tag{2.29}$$

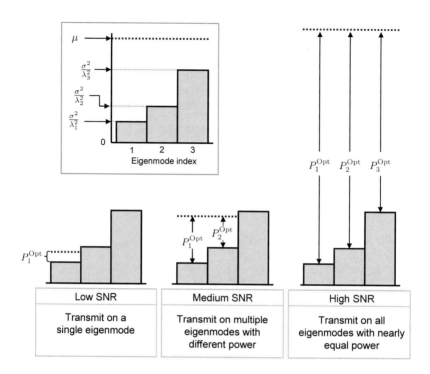

Fig. 2.3 The waterfilling algorithm determines the optimal allocation of power among parallel Gaussian channels that result from the decomposition of a MIMO channel. The channel has three eigenmodes, and the power allocation is shown for low, medium, and high values of SNR (P/σ^2).

(In the context of fast fading channels, the average open-loop capacity as defined here can also be interpreted as the "ergodic capacity" of the channel [39].) We will typically assume an i.i.d. Rayleigh distribution for the components of $\mathbf{H}$.

2.2.2.1 Low SNR

In the low-SNR regime where P approaches zero, the open-loop capacity for a fixed channel $\mathbf{H}$ with rank $r \leq \min(M, N)$ can be approximated as

$$C^{\mathrm{OL}}(\mathbf{H}, P/\sigma^2) = \log_2 \det \left(\mathbf{I}_N + \frac{P}{M\sigma^2} \mathbf{H}\mathbf{H}^H \right) \qquad (2.30)$$

$$= \log_2 \prod_{i=1}^{r} \left(1 + \frac{P}{M\sigma^2} \lambda_i^2(\mathbf{H}) \right) \tag{2.31}$$

$$\approx \log_2 \left(1 + \sum_{i=1}^{r} \frac{P}{M\sigma^2} \lambda_i^2(\mathbf{H}) \right) \tag{2.32}$$

$$= \log_2 \left[1 + \frac{P}{M\sigma^2} \operatorname{tr} \left(\mathbf{H}\mathbf{H}^H \right) \right] \tag{2.33}$$

$$\approx \frac{P}{M\sigma^2} \operatorname{tr} \left(\mathbf{H}\mathbf{H}^H \right) \log_2 e, \tag{2.34}$$

where $\lambda_i^2(\mathbf{H})$ are the eigenvalues of $\mathbf{H}\mathbf{H}^H$ (and $\lambda_i(\mathbf{H})$ are the singular values of $\mathbf{H}$). Equation (2.32) follows from the dominance of the linear terms for P approaching zero, and (2.34) follows from the approximation

$$\log_2(1 + x) \approx x \log_2 e \tag{2.35}$$

for x approaching zero. Therefore the average open-loop capacity for i.i.d. Rayleigh channels at low SNR is:

$$\bar{C}^{\mathrm{OL}}(M, N, P/\sigma^2) \approx \frac{P}{M\sigma^2} \mathbb{E}\left[\operatorname{tr}(\mathbf{H}\mathbf{H})^H \right] \log_2 e \tag{2.36}$$

$$= \frac{P}{M\sigma^2} \mathbb{E}\left[\sum_{m=1}^{M} \sum_{n=1}^{N} |h_{n,m}|^2 \right] \log_2 e \tag{2.37}$$

$$= N \frac{P}{\sigma^2} \log_2 e, \tag{2.38}$$

where (2.38) follows from $\mathbb{E}\left[\sum_{m=1}^{M} \sum_{n=1}^{N} |h_{n,m}|^2 \right] = MN$ for i.i.d. Rayleigh channels. Hence at low SNR, the average open-loop capacity scales linearly with the number of receive antennas N:

$$\lim_{P/\sigma^2 \to 0} \frac{\bar{C}^{(\mathrm{OL})}(M, N, P/\sigma^2)}{P/\sigma^2} = N \log_2 e. \tag{2.39}$$

In the low-SNR regime, multiple transmit antennas do not improve the capacity, and the capacity of any (M, N) channel with $M \geq 1$ is asymptotically equivalent.

For the closed-loop capacity, waterfilling at asymptotically low SNR puts all the power P into the single best eigenmode. (With i.i.d. Rayleigh channels, the singular values will be unique with probability 1.) The average capacity is therefore

$$\bar{C}^{(\mathrm{CL})}(M, N, P/\sigma^2) = \mathbb{E}\left[\max_{\sum P_i \leq P} \sum_{i=1}^{r} \log_2\left(1 + \frac{P_i}{\sigma^2}\lambda_i^2(\mathbf{H})\right)\right] \quad (2.40)$$

$$\approx \mathbb{E}\left[\max_{\sum P_i \leq P} \sum_{i=1}^{r} \frac{P_i}{\sigma^2}\lambda_i^2(\mathbf{H})\right]\log_2 e \quad (2.41)$$

$$\approx \frac{P}{\sigma^2}\mathbb{E}\left(\lambda_{\max}^2(\mathbf{H})\right)\log_2 e, \quad (2.42)$$

where (2.41) follows from (2.35), and $\lambda_{\max}^2(\mathbf{H})$ is the maximum eigenvalue value of $\mathbf{HH}^H$. Hence

$$\lim_{P/\sigma^2 \to 0} \frac{\bar{C}^{(\mathrm{CL})}(M, N, P/\sigma^2)}{P/\sigma^2} = \mathbb{E}\left(\lambda_{\max}^2(\mathbf{H})\right)\log_2 e. \quad (2.43)$$

Because $\mathbb{E}\left(\lambda_{\max}^2(\mathbf{H})\right) \geq \max(M, N)$ for i.i.d. Rayleigh channels, closed-loop MIMO capacity at low SNR benefits from combining at either the transmitter or receiver.

2.2.2.2 High SNR

From (2.20), the average capacity for i.i.d. Rayleigh channels in the limit of high SNR can be written as:

$$\bar{C}^{(\mathrm{OL})}(M, N, P/\sigma^2) = \mathbb{E}\left[\sum_{i=1}^{\min(M,N)} \log_2\left(1 + \frac{P}{M\sigma^2}\lambda_i^2(\mathbf{H})\right)\right]$$

$$\approx \min(M, N)\log_2\left(\frac{P}{M\sigma^2}\right) + \sum_{i=1}^{\min(M,N)} \mathbb{E}\left(\log_2 \lambda_i^2(\mathbf{H})\right), (2.44)$$

where (2.44) derives from the following approximation for large x:

$$\log_2(1 + x) \approx \log_2(x). \quad (2.45)$$

Because $\mathbb{E}\left(\log_2 \lambda_i^2(\mathbf{H})\right) > -\infty$ for all i, it follows that

$$\lim_{P/\sigma^2 \to \infty} \frac{\bar{C}^{\mathrm{OL}}(M, N, P/\sigma^2)}{\log_2 P/\sigma^2} = \min(M, N). \quad (2.46)$$

Therefore open-loop MIMO achieves a multiplexing gain of $\min(M, N)$ at high SNR.

For the closed-loop capacity at asymptotically high SNR, waterfilling puts equal power in each of the $\min(M, N)$ eigenmodes. Therefore the average capacity is

$$
\bar{C}^{(\mathrm{CL})}(M, N, P/\sigma^2) = \mathbb{E}\left[\max_{\sum P_i \le P} \sum_{i=1}^{\min(M,N)} \log_2\left(1 + \frac{P_i}{\sigma^2}\lambda_i^2(\mathbf{H})\right)\right]
$$

$$
= \mathbb{E}\left[\sum_{i=1}^{\min(M,N)} \log_2\left(1 + \frac{P/\sigma^2}{\min(M,N)}\lambda_i^2(\mathbf{H})\right)\right]
$$

$$
\approx \min(M,N)\log_2\left(\frac{P/\sigma^2}{\min(M,N)}\right) + \sum_{i=1}^{\min(M,N)} \mathbb{E}\left(\log_2\lambda_i^2(\mathbf{H})\right), \quad (2.47)
$$

where (2.47) follows from (2.45). Hence

$$
\lim_{P/\sigma^2 \to \infty} \frac{\bar{C}^{\mathrm{CL}}(M, N, P/\sigma^2)}{\log_2 P/\sigma^2} = \min(M, N), \quad (2.48)
$$

and closed-loop MIMO achieves the same multiplexing gain as open-loop MIMO (2.46) despite the advantage of CSIT. For both open- and closed-loop MIMO, the multiplexing gain at high SNR requires multiple antennas at both the transmitter and receiver.

2.2.2.3 Large number of antennas

When M and N go to infinity with the ratio M/N converging to α, and the SNR remains fixed at P, the open-loop capacity per transmit antenna converges almost surely to a constant [40]:

$$
\lim_{\substack{M,N \to \infty \\ M/N \to \alpha}} \frac{\bar{C}^{\mathrm{OL}}(M, N, P/\sigma^2)}{M} = \log\left[1 + S\left(\alpha, \frac{P/\sigma^2}{\alpha}\right)\right] + \frac{1}{P/\sigma^2}S\left(\alpha, \frac{P/\sigma^2}{\alpha}\right)
$$

$$
- \frac{1}{\alpha} + \frac{1}{\alpha}\log\left[1 + P/\sigma^2 - \frac{P/\sigma^2}{\alpha} + S\left(\alpha, \frac{P/\sigma^2}{\alpha}\right)\right], \quad (2.49)
$$

where

$$
S(\alpha, \rho) = \frac{1}{2}\left[\rho - \rho\alpha - 1 + \sqrt{(\rho - \rho\alpha - 1)^2 + 4\rho}\right].
$$

Thus the capacity grows linearly with the number of antennas. The quantity $S(\alpha, \rho)$ can be interpreted as the asymptotic SINR at the output of a linear MMSE receiver for the signal from each of the M transmit antennas.

For the open-loop $(M, 1)$ MISO channel with i.i.d. Rayleigh distribution, the transmit power is distributed among the M antennas, and the average capacity is given by

$$\bar{C}^{\mathrm{OL}}(M, 1, P/\sigma^2) = \mathbb{E}\left[\log_2\left(1 + \frac{P}{M\sigma^2}Z,\right)\right], \qquad (2.50)$$

where Z is a chi-square random variable with $2M$ degrees of freedom. Asymptotically, as $M \to \infty$, the open-loop MISO capacity converges as a result of the law of large numbers:

$$\lim_{M\to\infty} \bar{C}^{\mathrm{OL}}(M, 1, P/\sigma^2) = \log_2\left(1 + P/\sigma^2\right). \qquad (2.51)$$

Therefore the result of transmit diversity (diversity is the only phenomenon taking place in multi-antenna transmission with single-antenna reception) is to remove the effect of fading when enough transmit antennas are available.

2.2.3 Performance comparisons

The CDF of the open-loop (4,1) MISO and (1,4) SIMO capacities are shown in Figure 2.4 for i.i.d. Rayleigh channel realizations with SNR $P/\sigma^2 = 10$. The circles indicate the average capacities. For MISO channels, the average capacity increases as M increases, and the CDF becomes steeper, indicating there is less variation in the capacity as a result of diversity gain.

On the other hand, receiver combining for the SIMO channel results in both diversity gain and combining gain. Increasing the number of receive antennas results in a steeper CDF due to diversity and a shift with respect to the open-loop MISO curve due to combining gains. The performance of a $(1, N)$ SIMO channel is equivalent to that of a $(N, 1)$ closed-loop MISO channel.

Figure 2.5 shows the average capacity of various link configurations versus SNR. The $(4, 1)$ OL MISO capacity yields a small improvement over the SISO performance. The $(4, 1)$ CL, and $(1, 4)$ performance is better as a result of coherent transmitter or combining gain, but the slope of the capacity curve with respect to $\log_2 P/\sigma^2$ is the same as SISO's. At high SNR, the open-loop and closed-loop MIMO techniques achieve a multiplexing gain of 4, indicated by the slope of the capacity. For asymptotically high SNR, the open-loop and closed-loop capacities are equivalent, and there is already negligible difference for SNRs greater than 20 dB. For MIMO, every doubling (3 dB increase) in

SNR results in 4 bps/Hz of additional capacity. For SIMO or MISO, every doubling results in only 1 bps/Hz of additional capacity.

Figure 2.6 shows the average capacity of the same link configurations for a lower range of SNRs. At very low SNRs, the optimal transmission strategy benefits from diversity and combining but not from multiplexing. Compared to $(1, 4)$ SIMO, additional transmit antennas under $(4, 4)$ OL MIMO do not provide any benefit. Using knowledge of the channel at the transmitter, $(4, 4)$ CL MIMO achieves additional capacity by steering power in the direction of the channel's dominant eigenmode. To better visualize the relative gains due to multiple antennas compared to SISO, Figure 2.7 shows the ratios of the MIMO, SIMO, and MISO average capacities versus the SISO average capacity as a function of SNR. For $(4, 4)$ OL MIMO, the ratio is 4 for both low and high SNRs but dips below 4 in between.

For CL MIMO, the number of transmitted streams as determined by waterfilling depends on the SNR. Figure 2.8 shows the average number of transmitted streams (average number of eigenmodes with nonzero power) for different antenna configurations as a function of SNR. For SNRs below -15 dB, capacity is achieved by transmitting a single stream for all cases. As the SNR increases, the probability of transmitting multiple streams increases. For $(2,2)$ and $(4,4)$ multiplexing the maximum number of streams $\min(M, N)$ occurs with probability 1 for SNRs of at least 30 dB. For $(2,4)$, full multiplexing with probability 1 occurs for SNRs of at least 10 dB.

Figure 2.9 shows the average capacity versus the number of antennas M for (M, M) MIMO and $(1, M)$ SIMO. Because the MIMO capacity is roughly $M \log_2(P/\sigma^2)$ for high SNR, the slope of the curve versus M depends on the SNR.

2.3 Transceiver techniques

The previous section describes the theoretical capacity of MIMO links but only hints at the transceiver (transmitter and receiver) structure required to achieve those rates. In this section we discuss transceiver implementation for achieving open- and closed-loop capacity and a number of relevant suboptimal techniques, including linear receivers and space-time coding.

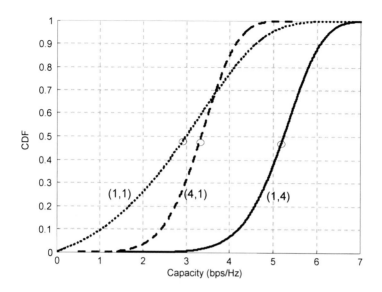

Fig. 2.4 CDF of capacity for SISO, open-loop MISO and SIMO channels for i.i.d. Rayleigh channels. The MISO channel increases reliability by providing diversity gain. The SIMO channel provides both diversity and combining gain.

2.3.1 Linear receivers

Let us consider the received signal (2.3) for the (M, N) MIMO channel where the data stream from the mth transmit antenna is highlighted:

$$\mathbf{x} = \mathbf{h}_m s_m + \sum_{j \neq m} \mathbf{h}_j s_j + \mathbf{n}, \tag{2.52}$$

where $\mathbf{h}_m$ is the mth column of the channel matrix $\mathbf{H}$. We assume that the power of the mth stream is $P_m := \mathbb{E}\left[|s_m|^2\right]$ and that the noise vector $\mathbf{n}$ is ZMSW Gaussian with covariance $\sigma^2 \mathbf{I}_M$.

We are interested in the class of *linear* receivers which computes a decision statistic r_m for the mth data stream by correlating the received signal $\mathbf{x}$ with an appropriately chosen vector $\mathbf{w}_m$:

$$r_m = \mathbf{w}_m^H \mathbf{x} \tag{2.53}$$

$$= (\mathbf{w}_m^H \mathbf{h}_m) s_m + \sum_{j \neq m} (\mathbf{w}_m^H \mathbf{h}_j) s_j + \mathbf{w}_m^H \mathbf{n}. \tag{2.54}$$

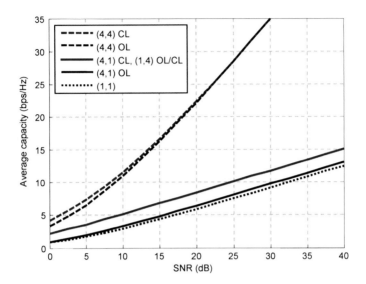

Fig. 2.5 Average capacity versus SNR for i.i.d. Rayleigh channels. At high SNR, (4,4) OL and CL MIMO provide a multiplexing gain of 4.

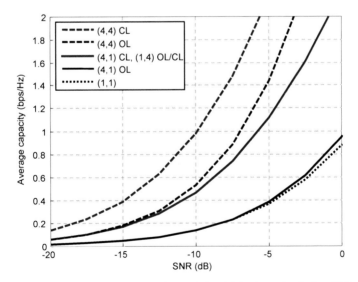

Fig. 2.6 Average capacity versus SNR for i.i.d. Rayleigh channels. At low SNR, (4,4)OL MIMO, (4,1)CL MISO, and (1,4)OL/CL SIMO provide a power gain of 4 as a result of combining.

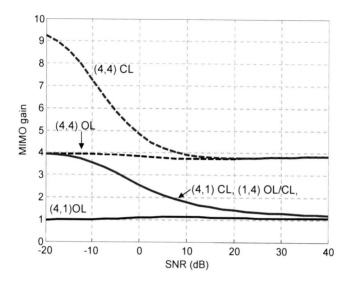

Fig. 2.7 Ratio of MIMO, SIMO, and MISO average rates versus the SISO average rate as a function of SNR for i.i.d. Rayleigh channels.

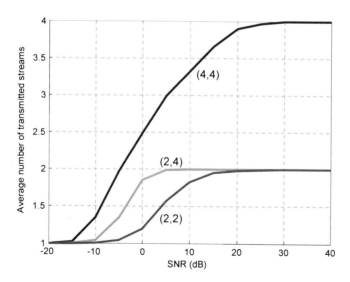

Fig. 2.8 Average number of transmitted streams for CL MIMO with i.i.d. Rayleigh channels. For lower SNRs, only a single stream is transmitted. Spatial multiplexing occurs at higher SNRs.

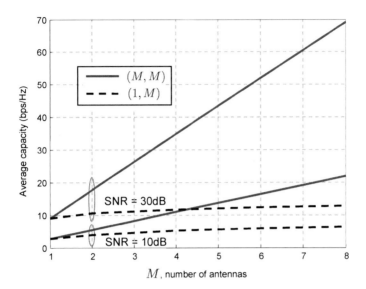

Fig. 2.9 Average capacity versus number of antennas M for i.i.d. Rayleigh channels. The slope of the MIMO capacity depends on the SNR.

The *output signal-to-noise ratio* (SNR), defined as the ratio of the receiver output power of the desired stream to the receiver output power of the thermal noise, is given by

$$\frac{\mathbb{E}\left[|\mathbf{w}_m^H \mathbf{h}_m s_m|^2\right]}{\mathbb{E}\left[|\mathbf{w}_m^H \mathbf{n}\right]} = \frac{|\mathbf{w}_m^H \mathbf{h}_m|^2 P_m}{||\mathbf{w}_m||^2 \sigma^2}. \tag{2.55}$$

In contrast to the SNR of the received signal P/σ^2 defined in Section 2.1, we emphasize that the output SNR (2.55) is defined at the output of the receiver processing. The *output signal-to-interference-plus-noise ratio* (SINR) is defined as the ratio of the receiver output power of the desired stream to the sum of the receiver output power of the thermal noise and interference from the other streams. Because the thermal noise and data streams are uncorrelated, the SINR is given by:

$$\frac{\mathbb{E}\left[\left|\mathbf{w}_m^H \mathbf{h}_m s_m\right|^2\right]}{\mathbb{E}\left[\left|\mathbf{w}_m^H \left(\sum_{j \neq m} \mathbf{h}_j s_j + \mathbf{n}\right)\right|^2\right]} = \frac{\left|\mathbf{w}_m^H \mathbf{h}_m\right|^2 P_m}{\sum_{j \neq m} \left|\mathbf{w}_m^H \mathbf{h}_j\right|^2 P_j + ||\mathbf{w}_m||^2 \sigma^2}. \tag{2.56}$$

We consider two linear receivers: the *matched filter* (MF) receiver and the *minimum mean-squared error* (MMSE) receiver. The MF is defined as the correlator matched to the desired stream's channel:

$$\mathbf{w}_m = \mathbf{h}_m. \tag{2.57}$$

Because the noise vector is Gaussian, this receiver maximizes the output SNR [41], but it is oblivious to the interference from the other data streams. It requires knowledge of the desired stream's channel but no knowledge of the other streams'. From (2.55), the output SNR is $\|\mathbf{h}_m\|^2 P_m/\sigma^2$, and the output SINR is

$$\Gamma_{\mathrm{MF},m} = \frac{\|\mathbf{h}_m\|^4 P_m}{\|\mathbf{h}_m\|^2 \sigma^2 + \sum_{j\neq m} |\mathbf{h}_m^* \mathbf{h}_j|^2 P_j}. \tag{2.58}$$

The MF receiver is also known as the *maximal ratio combiner* (MRC) because it weights and combines the received signal components to maximize the output SNR.

The MMSE receiver is a more sophisticated linear receiver that accounts for the presence of interference by minimizing the mean-squared error between the receiver output and the desired data stream s_m:

$$\begin{aligned}
\mathbf{w}_m &= \arg\min_{\mathbf{w}} \mathbb{E}\left[\left|\mathbf{w}^H \mathbf{x} - s_m\right|^2\right] \\
&= \arg\min_{\mathbf{w}} \mathbf{w}^H \left(\mathbf{HPH}^H + \sigma^2 \mathbf{I}_N\right) \mathbf{w} - 2\mathbf{w}^H \mathbf{h}_m P_m + P_m \\
&= \left(\mathbf{HPH}^H + \sigma^2 \mathbf{I}_N\right)^{-1} \mathbf{h}_m P_m,
\end{aligned}$$

where $\mathbf{P} := \mathrm{diag}(P_1,\ldots,P_M)$ is the diagonal matrix of powers. Using the matrix inversion lemma [42] for invertible $\mathbf{A}$:

$$\left(\mathbf{A} + \mathbf{bb}^H\right)^{-1} = \mathbf{A}^{-1} - \frac{\mathbf{A}^{-1}\mathbf{bb}^H \mathbf{A}^{-1}}{1 + +\mathbf{b}^H \mathbf{A}^{-1}\mathbf{b}}, \tag{2.59}$$

and defining $\mathbf{X} := \sum_{j\neq m} \mathbf{h}_j \mathbf{h}_j^H P_j + \sigma^2 \mathbf{I}_N$, we can write the MMSE receiver as

$$\begin{aligned}
\mathbf{w}_m &= \left(\mathbf{HPH}^H + \sigma^2 \mathbf{I}_N\right)^{-1} \mathbf{h}_m P_m \\
&= \left(\mathbf{X} + \mathbf{h}_m \mathbf{h}_m^H P_m\right)^{-1} \mathbf{h}_m P_m \\
&= \mathbf{X}^{-1} \mathbf{h}_m P_m - \frac{\mathbf{X}^{-1}\mathbf{h}_m \mathbf{h}_m^H \mathbf{X}^{-1}\mathbf{h}_m P_m^2}{1 + \mathbf{h}_m^H \mathbf{X}^{-1}\mathbf{h}_m P_m}
\end{aligned}$$

$$= \frac{\mathbf{X}^{-1}\mathbf{h}_m P_m}{1 + \mathbf{h}_m^H \mathbf{X}^{-1}\mathbf{h}_m P_m}. \tag{2.60}$$

Using (2.56) and (2.60), the SINR of the mth stream at the MMSE receiver output is

$$\Gamma_{\mathrm{MMSE},m} = \frac{\mathbf{w}_m^H \mathbf{h}_m P_m \mathbf{h}_m^H \mathbf{w}_m}{\mathbf{w}_m^H \mathbf{X} \mathbf{w}_m}$$

$$= \frac{P_m}{\sigma^2} \mathbf{h}_m^H \left(\mathbf{I}_N + \sum_{j \neq m} \frac{P_j}{\sigma^2} \mathbf{h}_j \mathbf{h}_j^H \right)^{-1} \mathbf{h}_m. \tag{2.61}$$

The MMSE receiver is the linear receiver which maximizes the SINR [43], and in this sense, it is often said to be the optimal linear receiver. We note that the MMSE receiver for a particular data stream can also be obtained by whitening the total noise plus interference affecting that stream, and then computing the matched filter for the equivalent channel after whitening.

As given in (2.60), the MMSE receiver requires knowledge of all channels. This requirement is reasonable for many situations where pilot signals are transmitted from each of the antennas. (See Section 2.7 for a discussion on acquiring channel estimates.) If the channel estimates are unreliable or unavailable, blind receivers techniques could be used [44].

Now suppose that M and N both go to infinity and $M/N \rightarrow \alpha$. In this large-system limit, it can be shown that [43]

$$\Gamma_{\mathrm{MF},m} \rightarrow \frac{P/\sigma^2}{\alpha(1 + P/\sigma^2)} \tag{2.62}$$

and

$$\Gamma_{\mathrm{MMSE},m} \rightarrow \frac{1}{2}\left[\left(\frac{P}{\sigma^2 \alpha} - \frac{P}{\sigma^2} - 1 \right) + \sqrt{\left(\frac{P}{\sigma^2 \alpha} - \frac{P}{\sigma^2} - 1 \right)^2 + \frac{4P}{\sigma^2 \alpha}} \right]. \tag{2.63}$$

Figure 2.10 shows the mean SINR (averaged over transmit antennas as well as i.i.d. Rayleigh channel realizations) versus average SNR P/σ^2 for both the MF and MMSE linear receivers. In each case, the solid lines show the results for $M = 4$ and $N = 4$ (as in (2.58) and (2.61)), while the dashed lines show the asymptotic results for $\alpha = 1$ (as in (2.62) and (2.63)). It can be observed that, in contrast to the interference-aware MMSE receiver, the SINR attainable with the interference-oblivious MF receiver saturates as the SNR is increased, indicating that it is interference-limited. (In fact, it can be

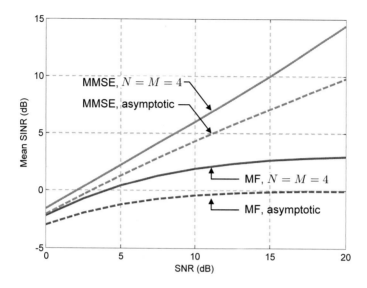

Fig. 2.10 Mean SINR per transmit antenna for MF and MMSE receivers. The asymptotic results assume that M and N both go to infinity with $M/N \to 1$.

shown [39] that the linear MMSE receiver attains the optimal multiplexing gain of $\min(M, N)$, i.e., the rate achievable with it as a function of SNR exhibits the same slope at high SNR as the capacity of the MIMO channel.) The trends exhibited by the asymptotic cases are already apparent for a link with relatively few antennas.

2.3.2 MMSE-SIC

The performance of the MMSE receiver could be improved by following it with a nonlinear *successive interference cancellation* (SIC) stage, shown in Figure 2.11. Suppose that we detect the data symbol s_1 from the first antenna. Its SINR is $\Gamma_{\mathrm{MMSE},1}$, given by (2.61). Assuming s_1 is detected correctly and assuming the receiver has ideal knowledge of $\mathbf{h}_1$, it can be cancelled from the received signal $\mathbf{x}$, yielding:

$$\mathbf{x} - \mathbf{h}_1 s_1 = \sum_{j=2}^{M} \mathbf{h}_j s_j + \mathbf{n}. \tag{2.64}$$

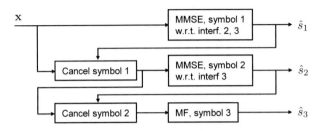

Fig. 2.11 MMSE-SIC detector for $M = 3$ symbols. Symbols are detected and cancelled in order, yielding estimates $\hat{s}_1, \hat{s}_2, \hat{s}_3$.

Given (2.64), data symbol s_2 can be detected using an MMSE receiver. Because there is no contribution from s_1 in (2.64), its SINR is:

$$\Gamma_2 = \frac{P_2}{\sigma^2} \mathbf{h}_2^H \left(\sum_{j=3}^{M} \mathbf{I}_N + \frac{P_j}{\sigma^2} \mathbf{h}_j \mathbf{h}_j^H \right)^{-1} \mathbf{h}_2.$$

If we successively detect the data symbols in order $s_3, s_4, \ldots, M$ and cancel their contributions, the mth stream ($m = 1, \ldots, M-1$) experiences interference from streams $m+1, m+2, \ldots, M$. The SINR for the mth stream is therefore

$$\Gamma_m = \frac{P_m}{\sigma^2} \mathbf{h}_m^H \left(\sum_{j=m+1}^{M} \mathbf{I}_N + \frac{P_j}{\sigma^2} \mathbf{h}_j \mathbf{h}_j^H \right)^{-1} \mathbf{h}_m. \tag{2.65}$$

The Mth stream is detected in the presence of only Gaussian noise. Using a matched filter, its SNR is

$$\Gamma_M = \frac{P_M}{\sigma^2} \|\mathbf{h}_M\|^2. \tag{2.66}$$

In the discussion of the MMSE-SIC detector (and the MF and MMSE detectors), we have focused on the detection of the data symbols s_m without any regard to channel encoding. In general, these data symbols are the output of a channel encoder, and we show in the next section how the MMSE-SIC detector in conjunction with a channel decoder can be used to achieve the open-loop MIMO capacity.

2.3.3 V-BLAST

The open-loop capacity for a fixed (M, N) MIMO link can be achieved using the *vertical BLAST* (V-BLAST) architecture, which uses an MMSE-SIC receiver structure where the interference cancellation is performed with respect to the decoded data streams. In the MIMO literature, the term "V-BLAST" is used to describe a variety of transceiver architectures and can be applied to block- or fast-fading channels. However, we use the term to refer specifically to the transmitter architecture shown in Figure 2.12 where the information data stream is multiplexed into M lower-rate streams that are independently encoded. This transmit architecture is sometimes referred to as *per-antenna rate control* (PARC) because the rate of each antenna's data stream is adjusted based on the channel realization $\mathbf{H}$.

From (2.3), the encoded transmitted signal vector for symbol time t is denoted as $\mathbf{s}^{(t)}$, and we let $\{\mathbf{s}^{(t)}\}$ denote the stream of vectors associated with a coding block. Similarly we let $\{\mathbf{x}^{(t)}\}$ denote the corresponding block of received signals. We assume that the channel is stationary for the duration of the coding block and that each stream has power $P_m = P/M$, $m = 1, \ldots, M$. Applying an MMSE detector to the received signal vectors $\{\mathbf{x}^{(t)}\}$, the output SINR is (2.61):

$$\Gamma_1 = \frac{P}{M\sigma^2} \mathbf{h}_1^H \left(\mathbf{I}_N + \sum_{j=2}^{M} \frac{P}{M\sigma^2} \mathbf{h}_j \mathbf{h}_j^H \right)^{-1} \mathbf{h}_1. \qquad (2.67)$$

If the data stream for the first antenna $\{s_1^{(t)}\}$ is encoded using a capacity-achieving code corresponding to rate $\log_2(1 + \Gamma_1)$, then it can be decoded without error. Using ideal knowledge of $\mathbf{h}_1$, its contribution to the received signal $\{\mathbf{x}^{(t)}\}$ can be cancelled. In general, if the mth data stream is encoded with rate $\log_2(1 + \Gamma_m)$, where from (2.65),

$$\Gamma_m = \frac{P}{M\sigma^2} \mathbf{h}_m^H \left(\mathbf{I}_N + \sum_{j=m+1}^{M} \frac{P}{M\sigma^2} \mathbf{h}_j \mathbf{h}_j^H \right)^{-1} \mathbf{h}_m, \qquad (2.68)$$

then it can be decoded and cancelled from the received signal so that data streams $m + 1, m + 2, \ldots, M$ do not experience interference from it. Using (2.68) and the matrix identities [42]

$$\frac{\det(\mathbf{A})}{\det(\mathbf{B})} = \det(\mathbf{B}^{-1}\mathbf{A}) \text{ and}$$

$$\det(\mathbf{I} + \mathbf{A}\mathbf{B}) = \det(\mathbf{I} + \mathbf{B}\mathbf{A}),$$

the rate achievable by stream m can be written as

$$\log_2(1 + \Gamma_m) = \log_2 \left[1 + \frac{P}{M\sigma^2}\mathbf{h}_m^H \left(\mathbf{I}_N + \sum_{j=m+1}^{M} \frac{P}{M\sigma^2}\mathbf{h}_j\mathbf{h}_j^H \right)^{-1} \mathbf{h}_m \right]$$

$$= \log_2 \det \left[\mathbf{I}_N + \frac{P}{M\sigma^2} \left(\mathbf{I}_N + \sum_{j=m+1}^{M} \frac{P}{M\sigma^2}\mathbf{h}_j\mathbf{h}_j^H \right)^{-1} \mathbf{h}_m\mathbf{h}_m^H \right]$$

$$= \log_2 \frac{\det \left(\mathbf{I}_N + \sum_{j=m}^{M} \frac{P}{M\sigma^2}\mathbf{h}_j\mathbf{h}_j^H \right)}{\det \left(\mathbf{I}_N + \sum_{j=m+1}^{M} \frac{P}{M\sigma^2}\mathbf{h}_j\mathbf{h}_j^H \right)}. \tag{2.69}$$

If we add the achievable rates for all M streams, then from (2.69), all the terms are cancelled except for the numerator of the rate for stream 1. The achievable sum rate is therefore

$$\sum_{m=1}^{M} \log_2(1 + \Gamma_m) = \log_2 \det \left(\mathbf{I}_N + \sum_{j=1}^{M} \frac{P}{M\sigma^2}\mathbf{h}_j\mathbf{h}_j^H \right)$$

$$= \log_2 \det \left(\mathbf{I}_N + \frac{P}{M\sigma^2}\mathbf{H}\mathbf{H}^H \right). \tag{2.70}$$

Noting the equivalence between (2.70) and (2.20), we conclude that the PARC strategy with an MMSE-SIC receiver achieves the open-loop MIMO capacity for block-fading channels.

For a fast-fading channel with i.i.d. Rayleigh distribution, using the statistics of $\mathbf{H}$, we can set the rate of stream m to be

$$R_m = \mathbb{E}_{\mathbf{H}} \left[\log_2 (1 + \Gamma_m) \right], \tag{2.71}$$

where Γ_m is from (2.68). Then, from (2.70), the achievable sum rate is

$$\sum_{m=1}^{M} \left[\mathbb{E}_{\mathbf{H}} \log_2(1 + \Gamma_m) \right] = \mathbb{E}_{\mathbf{H}} \left[\log_2 \det \left(\mathbf{I}_N + \frac{P}{M\sigma^2}\mathbf{H}\mathbf{H}^H \right) \right]. \tag{2.72}$$

Therefore the V-BLAST architecture also achieves the ergodic capacity for fast-fading channels. We emphasize that for a block-fading channel, the rates

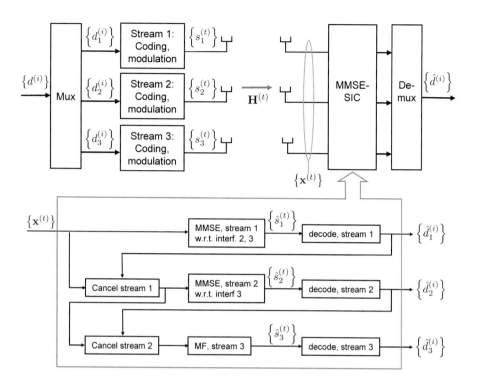

Fig. 2.12 Transceiver for achieving OL-MIMO capacity using the V-BLAST transmit architecture and an MMSE-SIC receiver.

are set as a function of the realization $\mathbf{H}$, while for a fast-fading channel, the rates are based on the statistics of $\mathbf{H}$.

Even though we have assumed that the data streams are decoded and cancelled in order from $m = 1$ to M, we note that the sum rate computed via (2.70) is independent of the order. Therefore, the OL MIMO capacity can be achieved for any ordering, as long as each antenna's stream is encoded with the appropriate rate as determined by the particular ordering.

The optimality of V-BLAST and PARC with MMSE-SIC was shown in [45] and is based on the sum-rate optimality of the MMSE-SIC for the multiple access channel [46]. This topic will be revisited in Chapter 3. While the PARC strategy assumes equal power on each stream, the throughput can be increased by optimizing the power distribution among the streams [45] using knowledge of $\mathbf{H}$ at the transmitter.

2.3.4 D-BLAST

In contrast to V-BLAST, where data streams for each antenna are encoded independently, *Diagonal BLAST* (D-BLAST) is an alternative technique for achieving the open-loop capacity that transmits the symbols for each coding block from all M antennas. Figure 2.13 shows the D-BLAST transceiver architecture. The information stream is encoded as U blocks of ML symbols. In the context of D-BLAST, each coding block is known as a *layer*. Layer $u = 1, \ldots, U$ consists of two subblocks of L symbols: $\left\{ b_1^{(u)} \right\}$ and $\left\{ b_2^{(u)} \right\}$. The blocks are transmitted in a staggered fashion so that L symbols $\left\{ b_2^{(u)} \right\}$ are transmitted from antenna 2, followed by L symbols $\left\{ b_1^{(u)} \right\}$ transmitted from antenna 1. The layer u transmission on antenna 1 occurs at the same time as the layer $u + 1$ transmission on antenna 2. During the first L symbol periods, symbols $\left\{ b_2^{(1)} \right\}$ are transmitted from antenna 2, and nothing is transmitted from antenna 1. During the last L symbol periods, $\left\{ b_1^{(U)} \right\}$ are transmitted from antenna 1, and nothing is transmitted from antenna 2.

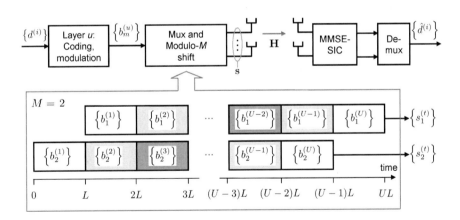

Fig. 2.13 The D-BLAST transmitter cyclicly shifts the association of each stream with all M antennas. The modulo-M shift is illustrated for $M = 2$ antennas and U layers.

To decode layer 1, symbols $\left\{ b_2^{(1)} \right\}$ are detected using a matched filter in the presence of thermal noise. The output SNR is

$$\Gamma_2 = \frac{P}{2\sigma^2}|\mathbf{h}_2|^2. \tag{2.73}$$

The symbols $\left\{b_1^{(1)}\right\}$ are detected using an MMSE receiver in the presence of interference from antenna 2 from symbols $\left\{b_2^{(2)}\right\}$. The output SINR is

$$\Gamma_1 = \frac{P}{2\sigma^2}\mathbf{h}_1^H \left(\mathbf{I}_N + \frac{P}{2\sigma^2}\mathbf{h}_2\mathbf{h}_2^H\right)^{-1}\mathbf{h}_1. \tag{2.74}$$

If the $2L$ symbols of layer 1 are encoded using a compound code at a rate $R < \log_2(1 + \Gamma_1) + \log_2(1 + \Gamma_2)$, then these symbols can be reliably decoded and canceled from the received signal stream.

The decoding of layer u follows the same procedure. Symbols $\left\{b_2^{(u)}\right\}$ are detected using a matched filter because symbols from the previous layer transmitted on antenna 1 have been cancelled. Then symbols $\left\{b_1^{(u)}\right\}$ are detected using an MMSE in the presence of interference from $\left\{b_2^{(u+1)}\right\}$. If U layers are transmitted, the achievable rate is

$$\frac{U}{U+1}\left[\log_2(1 + \Gamma_1) + \log_2(1 + \Gamma_2)\right], \tag{2.75}$$

where the fraction is due to the empty frames during the first and last layers. This overhead vanishes as U increases.

The procedure described for the $M = 2$ antenna case can be generalized for more antennas, so that the symbols for a single layer are staggered over ML symbol periods and transmitted over all M antennas. The symbols from antenna $m = 1, ..., M$ are detected in the presence of interference from antennas $m + 1, ..., M$. The SINR achieved for detecting symbols from antenna m is (2.68), and the overall achievable rate (for large U) is the open-loop MIMO capacity (2.70).

In practice, because the transmitter does not have knowledge of the channel, it does not know at what rate to encode the information. The receiver could estimate the channel $\mathbf{H}$, determine the channel capacity, and feed back this information to the transmitter. The D-BLAST encoder needs to know only the MIMO capacity whereas the V-BLAST encoder needs to know the achievable rates of each stream. D-BLAST would therefore require less feedback. However, due to the difficulty in implementing efficient compound codes, the V-BLAST architecture is more commonly implemented.

2.3.5 Closed-loop MIMO

If the channel $\mathbf{H}$ is known at the transmitter, one can achieve capacity by transmitting on the eigenmodes of the channel, as discussed in Section 2.2.1.1. The corresponding transceiver structure is shown in Figure 2.14. The information bit stream is first multiplexed into $\min(M, N)$ lower-rate streams, and the streams are encoded independently according to the rates determined by waterfilling. Given the SVD of the MIMO channel $\mathbf{H} = \mathbf{U}\mathbf{\Lambda}\mathbf{V}^H$, the $\min(M, N)$ streams are precoded using the first $\min(M, N)$ columns of the $M \times M$ unitary matrix $\mathbf{V}$. At the receiver, the $N \times N$ linear transformation $\mathbf{U}^H$ is applied, and the elements of the first $\min(M, N)$ rows are demodulated and decoded. The information bits for the $\min(M, N)$ streams are demultiplexed to create an estimate for the original bit stream.

For the special case of the $(M, 1)$ MISO channel, the data stream is precoded with the unit-normalized complex conjugate of the channel vector $\mathbf{h} \in \mathbb{C}^{1 \times M}$: ($\mathbf{v} = \mathbf{h}^H / \|\mathbf{h}\|$). This weighting is sometimes known as *maximal ratio transmission* (MRT), and it is the dual of MRC receiver for the $(1, N)$ SIMO channel.

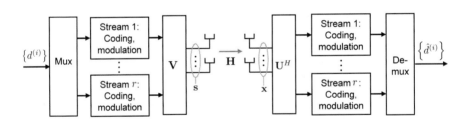

Fig. 2.14 Capacity-achieving transceiver for CL MIMO based on the SVD of $\mathbf{H}$. The power allocated to each stream is determined by waterfilling.

2.3.6 Space-time coding

If the channel is known at the transmitter, the SVD-based strategy described in Section 2.3.5 achieves the closed-loop capacity for any (M, N) link. If the channel is not known at the transmitter, the strategies for achieving open-loop capacity described in Section 2.3.5 apply only when $M \leq N$. *Space-time coding* is a class of techniques for achieving diversity gains in MISO channels when the channel state information is not known at the transmitter [47] [48]. Multiple receive antennas could be used to achieve combining and additional diversity gains. We will outline the basic principles of *space-time block codes* (STBCs), which are illustrative and representative of what space-time coding can achieve in MISO channels. The space-time block-coding transmission architecture is shown in Figure 2.15.

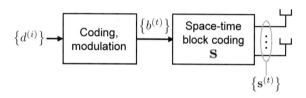

Fig. 2.15 In a space-time block-coding transmission architecture, the coded symbols $\{b^{(t)}\}$ are mapped to the transmitted symbol vector using, for example, (7.2)

In this architecture, a data stream is encoded using an outer channel encoder, and a space-time block encoder maps a block of Q encoded symbols $b_1, \ldots, b_Q$ onto the M antennas over L symbol periods. This mapping is represented by an $M \times L$ matrix $\mathbf{S}$, where the (m, l)th element of $\mathbf{S}$ ($m = 1, \ldots, M$, $l = 1, \ldots, L$) is the symbol transmitted from antenna m during symbol period l. In general, each element of $\mathbf{S}$ is a linear combination of $b_1, \ldots, b_Q$ and of the respective complex conjugates of these symbols $b_1^*, \ldots, b_Q^*$. The parameters L and $R := Q/L$ are known respectively as the *code delay* and *code rate* of space-time block code $\mathbf{S}$. Typically, the mapping parameters are chosen so that $L \geq M$ and $R \leq 1$. The performance of a space-time code can be measured by its *diversity order* which we define as the magnitude of the slope of the average symbol error rate at the receiver versus SNR (in a log-log scale).

For an $(M, 1)$ channel, the optimal (maximum) diversity order is M and can be achieved if $\mathbf{SS}^H$ is proportional to the identity matrix $\mathbf{I}_M$ [49]. It is also desirable for a code to be *full rate* (i.e., $R = 1$ and $Q = L$) and *delay optimal* (i.e., $L = M$) so that the code is time efficient. A space-time block code is *ideal* if it is full rate and delay optimal (so that $L = M = Q$) and if it achieves maximum diversity order.

For the case of $M = 2$ transmit antennas, a very popular and remarkably efficient space-time block code (the one that really defined the class of space-time block codes) is the *Alamouti space-time block code* [50]. Given a sequence of encoded symbols $\{b^{(t)}\}$ ($t = 0, 1, \ldots$), each pair of symbols on successive time intervals $b^{(2j)}$ and $b^{(2j+1)}$ ($j = 0, 1, \ldots$) is transmitted over the two antennas on intervals $2j$ and $2j + 1$ as follows:

$$\mathbf{s}^{(2j)} = \begin{bmatrix} b^{(2j)} \\ b^{(2j+1)} \end{bmatrix} \text{ and } \mathbf{s}^{(2j+1)} = \begin{bmatrix} -b^{*(2j+1)} \\ b^{*(2j)} \end{bmatrix}.$$

This code is ideal because it achieves maximum diversity order $M = 2$ with $L = M = Q = 2$. Moreover, it quite remarkably achieves the open-loop capacity for the (2,1) MISO channel for any SNR if an outer capacity-achieving scalar code is used [51] [52]. It also achieves the optimal diversity/ multiplexing tradeoff (see e.g. [53]). For a $(2, N)$ channel with $N > 1$, the Alamouti STBC with maximal ratio combining in general does not achieve the capacity. Not surprisingly, due to its remarkable properties, the Alamouti code has been used in several wireless standards.

To date, no ideal space-time block codes have been found for $M > 2$. However, quasi-orthogonal STBCs have also been proposed that approach the open-loop capacity in the (4,1) case [54] [55]. While generalizations of the quasi-orthogonal concept (and other space-time coding techniques) to arbitrary numbers of antennas have been suggested [56], it should be emphasized that the marginal diversity gains for open-loop MISO techniques diminish as the number of antennas increases. This fact, coupled with the additional overhead required to the channels, reduces the incentive for using too many ($M > 4$) antennas for space-time coding.

Besides STBCs, other classes of space-time codes include space-time trellis codes [47], linear dispersion codes [57], layered turbo codes [58], and lattice space-time codes [59].

2.3.7 Codebook precoding

If CSIT is ideally known, precoding with waterfilling achieves the closed-loop MIMO capacity. In many practical cases, it is not possible to obtain reliable

CSIT (see Section 4.4). Isotropic transmission is suboptimal, and as we saw in Section 2.2.3, the performance gap between OL-MIMO and CL-MIMO is significant at lower SNRs.

Another suboptimal alternative is to use precoding matrices that are chosen from a finite discrete set known as a *codebook*. (The precoding matrices are sometimes known as *codewords*, but they are not to be confused with the codewords associated with channel encoding.) The codebook is known by both the transmitter and receiver. Under codebook precoding, typically the receiver estimates the channel $\mathbf{H}$ and sends information back to the transmitter to indicate its preferred codeword. Using B feedback bits, the receiver can index up to 2^B codewords in the codebook. A block diagram is shown in Figure 2.16, where the codebook $\mathcal{B}$ consists of 2^B precoding matrices $\mathbf{G}_1, \ldots, \mathbf{G}_{2^B}$. This precoding technique is sometimes known as *limited feedback precoding*.

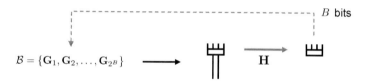

Fig. 2.16 Block diagram for codebook precoding. The user estimates the channel $\mathbf{H}$ and feeds back B bits to indicate its perfered precoding vector. The codebook $\mathcal{B}$ consists of 2^B codeword matrices and is known by the transmitter and the user.

In general, the transmitter can send up to $\min(M, N)$ data streams. If we let $J \leq \min(M, N)$ denote the number of streams, and $\mathbf{u} \in \mathbb{C}^J$ be the vector of data symbols, then the precoding matrix $\mathbf{G} \in \mathcal{B}$ is size $M \times J$, and the transmitted signal is given by $\mathbf{s} = \mathbf{G}\mathbf{u}$. From 2.3, the received signal is

$$\mathbf{x} = \mathbf{H}\mathbf{G}\mathbf{u} + \mathbf{n}. \tag{2.76}$$

When $J = 1$ and the input covariance has rank 1, precoding is often known as *beamforming*.

If the MIMO channel is changing sufficiently slowly, the mobile feedback could be aggregated over multiple feedback intervals so that the aggregated bits index a larger codebook. In general, a larger codebook implies more accurate knowledge of the MIMO channel at the transmitter, resulting in improved throughput. By aggregating the feedback bits over multiple intervals, the codewords can be arranged in a hierarchical tree structure so that the feedback on a given interval is an index of codewords that are the "children nodes" of a codeword indexed by previous feedback [60]. Temporal correlation of the channel can also be exploited by adapting codebooks over time [61] or by tracking the eigenmodes of the channel [62] [63] [64].

2.3.7.1 Single-antenna receiver, $N = 1$

Let us consider the problem of designing a codebook $\mathcal{B}$ for the case of a single-antenna receiver $N = 1$. In this case, the codebook consists of 2^B beamforming vectors: $\mathcal{B} = \{\mathbf{g}_1, \ldots, \mathbf{g}_{2^B}\}$, with $\mathbf{g}_b \in \mathbb{C}^{M \times 1}$. Assuming that the channel $\mathbf{h} \in \mathbb{C}^{1 \times M}$ can be estimated ideally, the user chooses the codeword in $\mathcal{B}$ which maximizes its rate:

$$\max_{\mathbf{g} \in \mathcal{B}} \log_2 \left(1 + |\mathbf{hg}|^2 \frac{P}{\sigma^2} \right) = \arg \max_{\mathbf{g} \in \mathcal{B}} |\mathbf{hg}|. \tag{2.77}$$

If the channel realizations are drawn from a finite, discrete distribution of 2^B M-dimensional vectors, one would design the codebook to consist of these vectors. The rate-maximizing codeword would be the (normalized) channel vector which corresponds to the maximal ratio transmitter (MRT). Assuming the channels could be estimated without error and the B feedback bits from each user could be received without error, the transmitter would achieve ideal CSIT. In practice, because the channel realizations are drawn from a continuous distribution, the codewords should be designed to optimally span the distribution, as determined by the channel correlation and desired performance metric.

At one extreme, the antennas are spatially uncorrelated, and the MISO channel coefficients each have an i.i.d. Rayleigh distribution. In this case the normalized realization $\mathbf{h}/\|\mathbf{h}\|$ is distributed uniformly on an M-dimensional unit hypersphere. The optimal rate maximizing strategy is to distribute the 2^B codewords as uniformly as possible on the surface of the hypersphere [65]. This problem is known as the Grassmannian line packing problem: design the codebook $\mathcal{B}$ to maximize the minimum distance between any two codewords

$$\sqrt{1 - \max_{i \neq j} |\mathbf{g}_i^H \mathbf{g}_j|^2}. \tag{2.78}$$

At the other extreme, the antennas are totally correlated, for example in a line-of-sight channel with zero angle spread. Figure 2.17 shows a linear array with M elements lying on the x-axis with uniform spacing d and a user with direction θ with respect to the x-axis. Let us consider the channel response $\mathbf{h}$ measured by a user lying in the general direction $\theta \in [0°, 360°)$. If the channel coefficient of the first element is $h_1 = \alpha \exp(j\gamma)$, then the coefficient at the mth element$(m = 1, ..., M)$ is

$$h_m(\theta) = \alpha \exp\left(\frac{2\pi j d}{\lambda}(m-1)\cos\theta + j\gamma\right), \qquad (2.79)$$

where λ is the carrier wavelength. We can use MRT to create a beamforming vector $\mathbf{g}(\theta^*)$ in the direction θ^* by matching the phase of the beamforming weight $g_m(\theta^*)$ to the phase of the channel coefficient $h_m(\theta^*)$, modulo the phase offset γ. With $d = \lambda/2$, the resulting MRT beamforming vector is

$$\mathbf{g}(\theta^*) = \frac{1}{\sqrt{M}} \begin{bmatrix} 1 \\ \exp(\pi j \cos\theta^*) \\ \vdots \\ \exp(\pi j (M-1)\cos\theta^*) \end{bmatrix}. \qquad (2.80)$$

Using this beamforming vector, the SNR of a user lying in the direction θ is

$$|\mathbf{h}^H(\theta)\mathbf{g}(\theta^*)|^2 P/\sigma^2. \qquad (2.81)$$

The MRT beamforming vector creates a *directional beam* (pointing in the direction θ^*) in the sense that the transmitted signal is co-phased to maximize the SNR of a user lying direction $\theta = \theta^*$. Figure 2.18 shows the MRT beam response for a linear array with $M = 4$ elements and a desired direction $\theta^* = 105°$. (The elements themselves are directional and pointing in the direction $\theta = 90°$, as described in Section 6.4.3. Otherwise, there would also be a response peak in the direction $\theta^* + 180°$.) Codewords could be designed to form directional beams spanning a desired range. For example, if users lie in a 120-degree sector $\theta \in [30°, 150°]$, we could choose to span this range using four MRT beams with directions $\{45°, 75°, 105°, 135°\}$. A user could determine its best codeword from (2.77) and indicate its preference with only $B = 2$ bits.

More general design techniques known as robust minimum variance beamforming can be used to design beamforming vectors for arbitrary antenna array configurations that are robust enough to withstand mismatch between

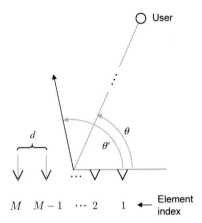

Fig. 2.17 An M-element linear array with inter-element spacing d. The direction of the user is θ, and a beam is pointed in the direction $\theta^* = 105°$.

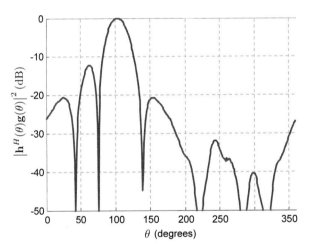

Fig. 2.18 The directional response as a function of the user direction θ for a MRT beam-forming vector (2.80) pointing in the direction $\theta^* = 105°$.

measured and actual channel state information $\mathbf{h}(\theta)$ [66]. The design of equal-gain beamformers with limited feedback [67] is also relevant for antenna arrays where the amplitudes of the channel coefficients are highly correlated.

For intermediate situations where the spatial channels are neither fully correlated nor totally uncorrelated, systematic codebook designs have been proposed in [68]. Codebooks can also be designed implicitly using a training sequence of channel realizations drawn from a given spatial correlation function. This technique, based on the Lloyd-Max algorithm [69], is effective for creating specifically tailored codebooks for arbitrary spatial correlations. The training sequence $\{\mathbf{H}_j\}_{j=1}^{N_{TS}}$ is of size N_{TS}, and its elements $\mathbf{H}_j$ are realizations of MIMO channels drawn from a given spatial correlation function. If we let $\mu(\mathbf{H}_j, \mathbf{g}_i)$ be a performance metric for a given channel realization and codebook vector, the algorithm iteratively maximizes the average performance metric

$$\max_{\mathcal{B}} \frac{1}{N_{TS}} \sum_{i=1}^{2^B} \sum_{\mathbf{H}_j \in \mathcal{R}_i} \mu(\mathbf{H}_j, \mathbf{g}_i), \qquad (2.82)$$

where $\mathcal{R}_i$ is the partitioned region of the training sequence associated with codeword $\mathbf{g}_i$. The size of the training sequence needs to scale at least linearly with the number of desired codewords to achieve good performance [69], hence the complexity of codebook design scales at least exponentially with the number of feedback bits B. However, because the codebook generation can be performed offline as long as the correlations are known beforehand, the complexity of the algorithm is not an issue. We also note that the algorithm converges to a maximum that is not guaranteed to be global. Nevertheless, it provides a practical way for codebook design even when the statistics of the source are not known or difficult to characterize.

2.3.7.2 Multi-antenna receiver, $N > 1$

If the receiver has multiple antennas $(N > 1)$, the beamforming techniques discussed for the case of single-antenna receivers could be used, and the received signal could be coherently combined across the N antennas. For i.i.d. Rayleigh channels with $M > 1$ and $N > 1$, the Grassmannian solution has been shown to maximize the beamforming rate [70] [71] [72].

In order to exploit the potential of spatial multiplexing, precoding matrices with rank $J > 1$ (and $J \leq \min(M, N)$) could be used. It is common to use *multidimensional eigenbeamforming*, where the columns of the precoding matrix $\mathbf{G} \in \mathbb{C}^{M \times J}$ are orthogonal such that $\mathbf{G}^H \mathbf{G}$ is a diagonal matrix. In doing so, the J streams are transmitted on mutually orthogonal subspaces, as is the case when precoding to achieve closed-loop MIMO capacity. We assume the symbols of the data vector $\mathbf{u} \in \mathbb{C}^J$ are independent and normalized such that $\mathbb{E}\left(\mathbf{u}\mathbf{u}^H\right) = \mathbf{I}_J$. Because the transmit power is $\mathrm{tr}\,\mathbb{E}\left(\mathbf{G}\mathbf{u}\mathbf{u}^H\mathbf{G}^H\right) = P$, we have that $\mathbf{G}^H\mathbf{G} = \mathrm{diag}(P_1, \ldots, P_J)$, where P_j $(j = 1, \ldots, J)$ is the power allocated to stream j, and $\sum_{j=1}^{J} P_j = P$.

Compared to transmitting with equal power on each stream, non-uniform power allocation requires more feedback and may not result in significant performance gains, especially if the channel is spatially uncorrelated. As described in Chapter 7, spatial multiplexing in 3GPP standards is achieved using codebook-based precoding with equal power on each stream.

A special case of multidimensional eigenbeamforming is to use antenna subset selection, where the columns of the precoder $\mathbf{G}$ are uniquely drawn from the columns of the $M \times M$ identity matrix and appropriately normalized [73]. In doing so, $J \leq \min(M, N)$ streams are uniquely associated with a subset of J transmit antennas. The case of $J = M$ corresponds to the V-BLAST transmission.

2.4 Practical considerations

2.4.1 CSI estimation

In deriving the MIMO capacity and capacity-achieving techniques, we have assumed that the CSI is known perfectly at the receiver and, when necessary for closed-loop MIMO, at the transmitter. In practice, estimates of the CSI at the receiver can be obtained from training signals (also known as pilot or reference signals) sent over time or frequency resources that are orthogonal to the data signals' resources. For M transmit antennas, the optimal training set consists of M mutually orthogonal signals, one assigned for each antenna and with equal power [74]. The reliability of the CSI estimates and the resulting rate performance depend on the fraction of resources devoted to the training signals and the rate of channel variation. As the channel varies more rapidly, additional training resources are required to achieve the same reliability in the channel estimates. (Reference signals are described in more detail in Section 4.4.)

To achieve closed-loop MIMO capacity, the CSI at the transmitter is assumed to be known ideally. If the CSIT is unreliable (see Section 4.4 for acquiring CSIT), the performance will be degraded. In high SNR channels, isotropic transmission should be used as an alternative because it requires no CSIT. In low SNR channels, precoding with limited feedback could be used to provide performance that is more robust to unreliable CSIT.

2.4.2 Spatial richness

The numerical results in this chapter assume that the MIMO channel coefficients are i.i.d. Rayleigh. As mentioned in Section 2.1, the spatial correlation between antennas depends on their spacing relative to the height of the surrounding scatterers. For a base station antenna that is high above the clutter (for example in rural or suburban deployments), a common rule of thumb is that spatial decorrelation could be achieved if the separation is at least 10 wavelengths [75]. On the other hand, if the base antenna is surrounded by scatterers of the same height (for example in rooftop urban deployments), decorrelation could be achieved with separation of only a few wavelengths. Decorrelation between pairs of base station antenna elements can also be achieved using cross-polarized antennas (Section 5.5.1). Mobile terminals are typically assumed to be surrounded by scatterers, and half-wavelength separation is considered sufficient for decorrelation [76].

Full multiplexing gain can be achieved over a SU-MIMO channel if the antenna array elements at both the transmitter and receiver are uncorrelated. As the antennas at either the transmitter or receiver become more correlated, the average capacity of the channel decreases. If the antennas at either end become fully correlated, then the channel cannot support multiplexing, and the multiplexing gain is 1. General characterizations of the capacity as a function of antenna correlation are given in [77].

2.5 Summary

Multiple-antenna techniques can be used to improve the throughput and reliability of wireless communication. In this chapter, we discussed the single-user (M, N) MIMO link where the transmitter is equipped with M antennas and the receiver is equipped with N antennas.

- The open-loop and closed-loop capacity measure the maximum rate of arbitrarily reliable communication for the case where the channel state information (CSI) is respectively known and not known at the transmitter. CSI at the receiver is always assumed.

- The MIMO capacity in a spatially rich channel scales linearly with the number of antennas. At high SNRs, a multiplexing gain of $\min(M, N)$ is achieved by transmitting multiple streams simultaneously from multiple antennas. At low SNRs, a power gain of N is achieved through receiver combining. CSIT for closed-loop MIMO allows more efficient power distribution, resulting in a higher capacity.

- Closed-loop MIMO capacity can be achieved using linear precoding and linear combining at the transmitter and receiver, respectively, where the transformations are based on the singular-value decomposition of the channel matrix **H**. Waterfilling is used to determine the optimal power allocation for each of the streams.

- Open-loop MIMO capacity for an (M, N) link (with $M \leq N$) can be achieved using isotropic transmission and an MMSE-SIC receiver. The V-BLAST transmit architecture achieves capacity by sending independent streams with appropriate rate assignment on each antenna. The D-BLAST transmit architecture achieves capacity by cyclically shifting the association of streams with transmit antennas.

- Space-time coding provides diversity gain for open-loop MISO channels. The Alamouti space-time block code achieves the capacity of a (2,1) MISO channel, but otherwise space-time coding cannot achieve the capacity of general MIMO antenna configurations. Space-time coding does not provide multiplexing gain and therefore provides only modest throughput gains as a result of diversity.

- In FDD systems where CSI at the transmitter is not readily available, feedback from the receiver can be used to index a fixed set of precoding matrices, providing suboptimal performance compared to the closed-loop MIMO capacity.

Chapter 3
Multiuser MIMO

Whereas in the previous chapter we considered a single-user MIMO link where data streams intended for a single user are multiplexed spatially, in this chapter we consider multiuser MIMO channels where data streams to/from multiple users are multiplexed in this way. In particular, we consider two multiuser Gaussian MIMO channels that are relevant for cellular networks. The uplink can be modeled by the Gaussian MIMO multiple-access channel (MAC), a many-to-one network where independent signals from $K > 1$ users are received by a single base receiver. The downlink can be modeled by the Gaussian MIMO broadcast channel (BC), a one-to-many network where a single base transmitter serves $K > 1$ users. To avoid needless repetition, we will henceforth drop the "Gaussian MIMO" qualifier when referring to these channels.

This chapter focuses on the K-dimensional capacity regions of the MAC and BC and describes techniques for achieving optimal rate vectors that maximize a weighted sum-rate metric. As we will see in Chapter 5, this metric is relevant for scheduling and resource allocation in cellular networks.

3.1 Channel models

The channel models for the MAC and BC were given in Chapter 1, but we repeat them here for convenience. A $((K, N), M)$ MAC has K users, each with N antennas, whose signals are received by a base with M antennas. The baseband received signal is

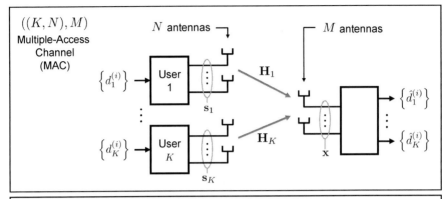

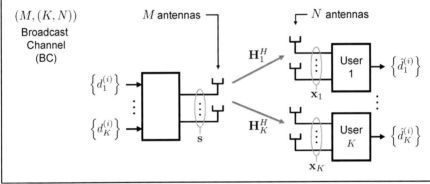

Fig. 3.1 $((K, N), M)$ multiple-access channel. K users, each with N antennas, transmit signals that are demodulated by a receiver with M antennas. $(M, (K, N))$ broadcast channel. Transmitter with M antennas serves K users, each with N antennas

$$\mathbf{x} = \sum_{k=1}^{K} \mathbf{H}_k \mathbf{s}_k + \mathbf{n}, \qquad (3.1)$$

where

- $\mathbf{x} \in \mathbb{C}^{M \times 1}$ is the received signal whose mth element $(m = 1, \ldots, M)$ is associated with antenna m
- $\mathbf{H}_k \in \mathbb{C}^{M \times N}$ is the channel matrix for the kth user $(k = 1, \ldots, K)$ whose (m, n)th entry $(m = 1, \ldots, M; n = 1, \ldots, N)$ gives the complex amplitude between the nth transmit antenna of user k and the mth receive antenna

- $\mathbf{s}_k \in \mathbb{C}^{N \times 1}$ is the transmitted signal vector from user k. The covariance is $\mathbf{Q}_k := \mathbb{E}\left(\mathbf{s}_k \mathbf{s}_k^H\right)$, and the signal is subject to the power constraint $\operatorname{tr} \mathbf{Q}_k \leq P_k$ with $P_k \geq 0$
- $\mathbf{n} \in \mathbb{C}^{M \times 1}$ is the additive noise and is a zero-mean spatially white (ZMSW) additive Gaussian random vector with variance σ^2

The components of each transmitted signal vector for the kth user $\mathbf{s}_k$ ($k = 1, \ldots, K$) are functions of the information bit stream for the kth user $\left\{d_k^{(i)}\right\}$. The receiver demodulates $\mathbf{x}$ and provides estimates for these data streams $\left\{\hat{d}_k^{(i)}\right\}$.

A $(M, (K, N))$ BC denotes a base station with M transmit antennas serving K users, each with N antennas. The base transmits a common signal $\mathbf{s}$ which contains the encoded symbols of the K users' data streams. This signal travels over the channel $\mathbf{H}_k$ to reach user k ($k = 1, \ldots, K$). The baseband received signal by the kth user can be written as:

$$\mathbf{x}_k = \mathbf{H}_k^H \mathbf{s} + \mathbf{n}_k, \tag{3.2}$$

where

- $\mathbf{x}_k \in \mathbb{C}^{N \times 1}$ whose nth element ($n = 1, \ldots, N$) is associated with antenna n of user k
- $\mathbf{H}_k^H \in \mathbb{C}^{N \times M}$ is the channel matrix for the kth user ($k = 1, \ldots, K$) whose (n, m)th entry ($n = 1, \ldots, N; m = 1, \ldots, M$) gives the complex amplitude between the nth transmit antenna of user k and the mth receive antenna
- $\mathbf{s} \in \mathbb{C}^{M \times 1}$ is the transmitted signal vector which is a function of the data signals for the K users. The covariance is $\mathbf{Q} := \mathbb{E}\left(\mathbf{s}\mathbf{s}^H\right)$, and the signal is subject to the power constraint $\operatorname{tr} \mathbf{Q} \leq P$, with $P \geq 0$
- $\mathbf{n}_k \in \mathbb{C}^{N \times 1}$ is the additive noise at user k and is a zero-mean spatially white (ZMSW) additive Gaussian random vector with variance σ^2

The components of the transmitted signal $\mathbf{s}$ are functions of the K users' information bit streams $\left\{d_k^{(i)}\right\}$ for $k = 1, \ldots, K$. The kth user provides estimates for its data stream $\left\{\hat{d}_k^{(i)}\right\}$ given its received signal $\mathbf{x}_k$.

3.2 Multiple-access channel (MAC) capacity region

For a $((K, N), M)$ MAC, we define a K-dimensional *capacity region*

$$\mathcal{C}^{\mathrm{MAC}}\left(\mathbf{H}_1, ..., \mathbf{H}_K, \frac{P_1}{\sigma^2}, ..., \frac{P_K}{\sigma^2}\right) \qquad (3.3)$$

consisting of achievable rate vectors $(R_1, \ldots, R_K)$, where $R_k \geq 0$ for $k = 1, \ldots, K$. For the point-to-point single-user channel, the capacity C gives the fundamental performance limit, in that reliable communication is possible for any rate $R < C$ and impossible for any rate $R > C$. For the MAC capacity region, any vector that belongs to the region $\mathcal{C}^{\mathrm{MAC}}$ is a K-tuple of rates that can be *simultaneously* achieved by the users. An important scalar performance metric derived from the capacity region is the sum-rate capacity (or sum-capacity), which is the maximum total throughput that can be achieved:

$$\max_{\mathbf{R} \in \mathcal{C}^{\mathrm{MAC}}} \sum_{k=1}^{K} R_k. \qquad (3.4)$$

In the remainder of this section, we characterize the capacity region for the MIMO MAC channel (3.1), starting with the single-antenna MAC. We describe how to achieve rate vectors lying on the boundary of the capacity region. A formal derivation of the MAC capacity region is beyond the scope of this book, but those interested can refer to [37].

3.2.1 Single-antenna transmitters ($N = 1$)

We consider the special case of (3.1) for the $((2,1),1)$ MAC channel where signals from $K = 2$ mobiles, each with a single antenna, are received by a base with a single antenna ($M = 1$). We let h_k ($k = 1, 2$) be the time-invariant SISO channel for each user, and we assume that these values are known at the base. The capacity region, shown in Figure 3.2 for fixed channel realizations h_1 and h_2, is the set of non-negative rate pairs (R_1, R_2) that satisfies the following three constraints:

$$R_1 \leq \log_2\left(1 + \frac{P_1}{\sigma^2}|h_1|^2\right)$$
$$R_2 \leq \log_2\left(1 + \frac{P_2}{\sigma^2}|h_2|^2\right)$$
$$R_1 + R_2 \leq \log_2\left(1 + \frac{P_1}{\sigma^2}|h_1|^2 + \frac{P_2}{\sigma^2}|h_2|^2\right). \qquad (3.5)$$

Figure 3.2 illustrates this capacity region. Point A on the capacity region boundary can be achieved by having user 1 transmit with a Gaussian codeword with power P_1 and having user 2 be silent. Similarly, point B is achieved with user 2 on and user 1 off. Each rate pair on the line segment joining A and B can be achieved by time-multiplexing the transmissions between the two users with an appropriate fraction of time for each user. The triangular region bounded by this segment is called the time-division multiple access (TDMA) rate region.

While time-multiplexing of users for the MAC seems like a reasonable strategy for achieving high throughput, it turns out we can do better by having the users transmit simultaneously. For example, point C on the boundary of the MAC capacity region can be achieved by having both users transmit with full power. The signal s_2 for user 2 is decoded in the presence of interference from user 1's signal and thermal noise. The rate achieved by user 2 is

$$R_2 = \log_2\left(1 + \frac{P_2|h_2|^2}{\sigma^2 + P_1|h_1|^2}\right). \tag{3.6}$$

Since h_2 and s_2 are known at the receiver, the received data signal for user 2 can then be reconstructed and subtracted from the received signal. The signal for user 1 can then be decoded in the presence of only thermal noise, achieving rate

$$R_1 = \log_2\left(1 + \frac{P_1}{\sigma^2}|h_1|^2\right). \tag{3.7}$$

Adding the rates from (3.6) and (3.7), the sum rate is:

$$R_1 + R_2 = \log_2\left(1 + \frac{P_1}{\sigma^2}|h_1|^2 + \frac{P_2}{\sigma^2}|h_2|^2\right),$$

thereby achieving the sum-rate capacity (3.5). By switching the decoding order, one can achieve the same sum rate but at point D. To summarize, the sum-rate capacity is achieved by having both users transmit with full power and using interference cancelation at the receiver. The order of the cancelation does not affect the sum rate, and any point between the vertices C and D can be achieved with time-sharing between the two decoding orders.

Generalizing to the K-user case, the K-dimensional capacity region for the $((K,1),1)$ MAC is bounded by $2^K - 1$ hyperplanes, each corresponding to a nonempty subset of users. The capacity region is a generalized pentagonal region known as a polymatroid [78]. The sum-rate capacity is achieved by having all users transmit with full power and using successive interference cancelation, where users are successively detected and canceled one by one in the presence of noise and interference from uncanceled users. As in the 2-user case, the sum-rate capacity does not depend on the cancelation order. In general, the sum-rate capacity region has $K!$ corner points, and these are each achieved using a different cancelation order. Generalizing further to multiple receive antennas $(M > 1)$, the $((K,1), M)$ MAC capacity region is

$$
\mathcal{C}^{\mathrm{MAC}}(\mathbf{h}_1, \ldots, \mathbf{h}_K, \tfrac{P_1}{\sigma^2}, \ldots, \tfrac{P_K}{\sigma^2}) =
$$
$$
\left\{ (R_1, \ldots, R_K) : \sum_{k \in S} R_k \leq \log_2 \det \left(\mathbf{I}_M + \sum_{k \in S} \tfrac{P_k}{\sigma^2} \mathbf{h}_k \mathbf{h}_k^H \right), \forall S \subseteq \{1, \ldots, K\} \right\},
$$
$$(3.8)$$

where there is one rate constraint for each nonempty subset S of users. Note that if the transmit powers of the users are all equal $(P_1 = P_2 = \ldots = P_K = P/K)$, then the sum-capacity of the $((K,1), M)$ MAC is equivalent to the (K, M) SU-MIMO open-loop capacity with power constraint P where the columns of the $M \times K$ channel are the K user's channels $(\mathbf{H} = [\mathbf{h}_1, \ldots, \mathbf{h}_K])$. Therefore, since the MMSE-SIC receiver achieves the SU-MIMO open-loop capacity (Section 2.3.3), it can be used to achieve the sum-capacity of the $((K,1), M)$ MAC where the users transmit with equal power. More generally, the MMSE-SIC receiver can achieve the MAC sum-capacity for any set of powers $P_1, \ldots, P_K$.

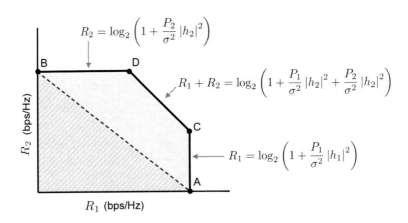

Fig. 3.2 $((2,1),1)$ MAC capacity region: $K = 2$ users, each with $N = 1$ antenna, and a receiver with $M = 1$ antenna. The TDMA rate region is bounded by the line segment joining points A and B.

3.2.2 Multiple-antenna transmitters (N > 1)

If the users have multiple antennas $(N > 1)$, the transmitted signal from user k is an N-dimensional vector $\mathbf{s}_k \in \mathbb{C}^{N \times 1}$ with covariance $\mathbf{Q}_k$. Generalizing the capacity region of a $((K, 1), M)$ MAC (3.8), the capacity region of a $((K, N), M)$ MAC with fixed covariances $\mathbf{Q}_1, \ldots, \mathbf{Q}_K$ is

$$
\mathcal{C}^{\mathrm{MAC}}(\{\mathbf{H}_k\}, \{\mathbf{Q}_k/\sigma^2\}) =
\left\{
\begin{array}{l}
(R_1, \ldots, R_K) : \\
\sum_{k \in S} R_k \leq \log_2 \det \left(\mathbf{I}_M + \frac{1}{\sigma^2} \sum_{k \in S} \mathbf{H}_k \mathbf{Q}_k \mathbf{H}_k^H \right), \forall S \subseteq \{1, ..., K\}
\end{array}
\right\}.
\tag{3.9}
$$

The shaded pentagonal region of Figure 3.3 shows an example of the capacity region for a $((2, N), M)$ MAC with fixed covariances $\mathbf{Q}_1$ and $\mathbf{Q}_2$, each with rank $r = 2$. The transmitted signal for user k $(k = 1, 2)$ can be written as $\mathbf{s}_k = \mathbf{G}_k \mathbf{u}_k$, where $\mathbf{G}_k \in \mathbb{C}^{N \times r}$ is the precoding matrix, and $\mathbf{u}_k \in \mathbb{C}^{r \times 1}$ is the data vector. Point A on the capacity region boundary can be achieved using an MMSE-SIC as depicted in Figure 3.4 where the streams of user 1 are detected first. By treating the transmitted signal of user k as a linear combination of $r = 2$ independent data streams, the $((2, N), M)$ MAC is equivalent to a MAC with four virtual users, each transmitting a single stream. The rates of the two users at point A are respectively

$$
R_1 = \log_2 \frac{\det \left(\mathbf{I}_M + \frac{1}{\sigma^2} \mathbf{H}_1 \mathbf{Q}_1 \mathbf{H}_1^H + \frac{1}{\sigma^2} \mathbf{H}_2 \mathbf{Q}_2 \mathbf{H}_2^H \right)}{\det \left(\mathbf{I}_M + \frac{1}{\sigma^2} \mathbf{H}_2 \mathbf{Q}_2 \mathbf{H}_2^H \right)}
$$

and

$$
R_2 = \log_2 \det \left(\mathbf{I}_M + \frac{1}{\sigma^2} \mathbf{H}_2 \mathbf{Q}_2 \mathbf{H}_2^H \right).
$$

For a different set of covariances $\mathbf{Q}_1$ and $\mathbf{Q}_2$ (subject to the power constraints $\mathrm{tr}\,\mathbf{Q}_1 = P_1$ and $\mathrm{tr}\,\mathbf{Q}_2 = P_2$), the capacity region will correspond to a different pentagonal region as indicated by the dashed boundaries in Figure 3.3. Taking the union of all such capacity regions yields the $((2, N), M)$ MAC capacity region $\mathcal{C}^{\mathrm{MAC}}(\mathbf{H}_1, \mathbf{H}_2, P_1/\sigma^2, P_2/\sigma^2)$ whose boundary is indicated by the heavy solid line in Figure 3.3.

In general, the $((K, N), M)$ MAC capacity region can be written as the union of capacity regions (3.9) with different covariances:

$$\mathcal{C}^{\text{MAC}}(\{\mathbf{H}_k\}, \{P_k/\sigma^2\}) =$$

$$\bigcup_{\substack{\mathbf{Q}_k \geq 0 \\ tr\mathbf{Q}_k \leq P_k}} \left\{ \begin{array}{l} (R_1, \ldots, R_K) : \\[4pt] \displaystyle\sum_{k \in S} R_k \leq \log_2 \det\left(\mathbf{I}_M + \frac{1}{\sigma^2}\sum_{k \in S}\mathbf{H}_k\mathbf{Q}_k\mathbf{H}_k^H\right), \forall S \subseteq \{1, \ldots, K\} \end{array} \right\}.$$

$$(3.10)$$

A capacity-achieving architecture for the MAC is shown in Figure 3.5. User k demultiplexes its information bits $\left\{d_k^{(i)}\right\}$ into $r_k \geq 1$ parallel streams, and then encodes and modulates each of these streams. The symbol vector $\mathbf{u}_k \in \mathbb{C}^{r_k}$ is weighted by a precoding matrix $\mathbf{G}_k \in \mathbb{C}^{M \times r_k}$ such that the transmitted signal is $\mathbf{s}_k = \mathbf{G}_k\mathbf{u}_k$. How are the transmit covariances $\mathbf{Q}_1, \ldots, \mathbf{Q}_K$ determined to achieve a particular rate vector on the capacity region boundary? Recall that for the single-user MIMO channel, the capacity-achieving covariance $\mathbf{Q}$ is based on the SVD of the MIMO channel $\mathbf{H}$. For the MAC, if all the users' MIMO channels are mutually orthogonal (i.e., $\mathbf{H}_i^H\mathbf{H}_j = \mathbf{0}_M$, for all $i \neq j$), then the capacity-achieving covariances for each user individually also achieve the MAC capacity region boundary. In general, however, because of the interaction of the users' MIMO channels, the MAC capacity region is not achieved using the optimal single-user covariances. Determining the capacity-achieving covariances for the MAC is not easy because of the channel interactions. While there is no known general closed-form expression for computing the optimal covariances for achieving the capacity region boundary, one can compute them numerically using iterative techniques described in Section 3.5.1.

For a given set of covariances $\mathbf{Q}_1, \ldots, \mathbf{Q}_K$ and a given decoding order for the users, the MMSE-SIC receiver successively decodes and cancels users so that each user's SINR is maximized in the presence of noise and interference from all other users not yet decoded. Without loss of generality, we can assume that users are decoded in order from user 1 up to user K. If the cancellation is ideal, then in decoding the data stream for user k ($k = 1, \ldots, K$), the received signal $\mathbf{x}$ (3.1) has been cleansed of the interference from users $1, \ldots, K-1$. Its virtual received signal is

$$\mathbf{x} - \sum_{j=1}^{k-1}\mathbf{H}_j\mathbf{s}_j = \sum_{j=k}^{K}\mathbf{H}_j\mathbf{s}_j + \mathbf{n}, \qquad (3.11)$$

so $\mathbf{s}_k$ is received in the presence of interference from users $k+1, \ldots, K$. Proceeding as in Section 2.3.2, we can see that user k achieves rate

$$R_k = \log_2 \frac{\det\left(\mathbf{I}_M + \frac{1}{\sigma^2}\sum_{j=k}^{K}\mathbf{H}_j\mathbf{Q}_j\mathbf{H}_j^H\right)}{\det\left(\mathbf{I}_M + \frac{1}{\sigma^2}\sum_{j=k+1}^{K}\mathbf{H}_j\mathbf{Q}_j\mathbf{H}_j^H\right)} \tag{3.12}$$

$$= \log_2\det\left[\mathbf{I}_M + \left(\sigma^2\mathbf{I}_M + \sum_{j=k+1}^{K}\mathbf{H}_j\mathbf{Q}_j\mathbf{H}_j^H\right)^{-1}\mathbf{H}_k\mathbf{Q}_k\mathbf{H}_k^H\right] \tag{3.13}$$

by whitening its virtual received signal (3.11) with respect to the noise and interference covariance and then performing MMSE-SIC for its desired symbol vector $\mathbf{u}_k$.

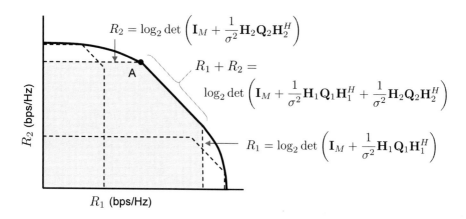

Fig. 3.3 The shaded pentagonal rate region is for a $K = 2$ MAC with fixed covariances $\mathbf{Q}_1$ and $\mathbf{Q}_2$, with $\operatorname{tr}\mathbf{Q}_k \leq P_k$. The general MAC capacity region is the union of all such pentagonal rate regions for different covariances that satisfy the power constraints.

3.3 Broadcast channel (BC) capacity region

For a $(M,(K,N))$ BC, we define the K-dimensional capacity region

$$\mathcal{C}^{\mathrm{BC}}\left(\mathbf{H}_1^H, ..., \mathbf{H}_K^H, P/\sigma^2\right) \tag{3.14}$$

as the set of rate vectors $(R_1,\ldots,R_K)$ that can be simultaneously achieved by the users. Whereas in the MAC the power of each user is individually constrained, in the BC, the transmitter is able to partition power among the

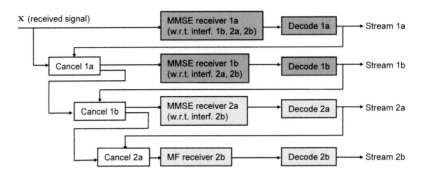

Fig. 3.4 An illustration of the MMSE with successive interference cancellation for achieving point A in the 2-user rate region of Figure 3.3

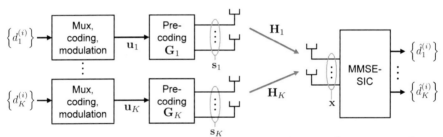

Fig. 3.5 A capacity-achieving strategy for the MIMO MAC. User k uses matrix $\mathbf{G}_k$ to precode its data vector $\mathbf{u}_k$. A MMSE-SIC is used at the receiver to jointly detect and demodulate the user data streams.

multiple users so the total power allocated to all users is constrained by P. In this section, we characterize the BC capacity region and discuss capacity-achieving techniques. As we will see, rate vectors on the BC capacity region boundary can be achieved using a transmitter encoding technique that is analogous to the capacity-achieving successive interference cancellation used for the MAC.

3.3.1 Single-antenna transmitters $(M = 1)$

We first consider the case of a $(1, (2, N))$ broadcast channel with a single-antenna transmitter serving two users each with $N \geq 1$ antennas. Suppose the $M \times 1$ SIMO channels $\mathbf{h}_1^H$ and $\mathbf{h}_2^H$ are fixed such that $||\mathbf{h}_1|| < ||\mathbf{h}_2||$. For $M = 1$, the broadcast channel is *degraded*, meaning that the users' channels can be ordered so that the received signal at a user can be regarded as a more noisy version of the signal received by the previous user in the ordering. In this case, we can imagine that user 1 is farther from the transmitter than user 2 and that it receives a further degraded version of the signal transmitted with power P. The capacity region of the 2-user degraded broadcast channel is given by the rate pairs (R_1, R_2) such that

$$R_1 \leq \log_2 \left(1 + \frac{P_1 ||\mathbf{h}_1||^2}{\sigma^2 + P_2 ||\mathbf{h}_1||^2} \right) \qquad (3.15)$$

$$R_2 \leq \log_2 \left(1 + \frac{P_2}{\sigma^2} ||\mathbf{h}_2||^2 \right), \qquad (3.16)$$

where P_1 and P_2 are the non-zero powers allocated to the two users that satisfy the power constraint $P_1 + P_2 \leq P$.

The rate vector corresponding to a given power allocation P_1 and P_2 can be achieved by encoding the data signals u_1 and u_2 for the users with respective powers $P_1 = \mathbb{E}\left(|u_1|^2\right)$ and $P_2 = \mathbb{E}\left(|u_1|^2\right)$. The transmitted signal is the superposition or sum of the codewords, $s = u_1 + u_2$. User 1 detects its signal in the presence of noise and interference u_2 to achieve rate R_1 given by (3.15). Because user 2 has a higher channel gain than user 1, it could detect u_1, cancel it, and detect u_2 in the presence of only AWGN.

Whereas superposition coding as above relies on interference cancelation at the receiver, an alternative technique for achieving the capacity region places the burden of complexity on the transmitter. This technique, known as *dirty paper coding* (DPC) [79], is an encoding process that occurs jointly and in an ordered fashion among the users so that a given user experiences interference only from users encoded after it. In other words, a given user receives zero power from the signals of users that are encoded earlier in the sequence, as if the signals of these users were "pre-subtracted" at the transmitter. In our example, user 1 would be encoded in a conventional (capacity-achieving) manner, then user 2 would be encoded using DPC with non-causal knowledge of user 1's data signal.

The left subfigure of Figure 3.6 shows the 2-user capacity region for $||\mathbf{h}_1|| <$ $||\mathbf{h}_2||$, obtained from (3.15) and (3.16) by varying the power partition between the users. We note that by switching the DPC encoding order so that user 2 is encoded first, user 1 is decoded without interference, and the resulting rate region is concave.

For the general K-user degraded broadcast channel (still with a single-antenna transmitter), we assume without loss of generality that the channels are ordered $||\mathbf{h}_1|| \leq ||\mathbf{h}_2|| \leq \ldots \leq ||\mathbf{h}_K||$. Under superposition coding, the kth user detects and cancels the signals from users $1, 2, \ldots, k-1$ whose channels are weaker. Under DPC, the users are encoded in order from user 1 to 2 up to K. The rate region is given by the rate vectors $R_1, \ldots, R_K$ satisfying

$$\mathcal{C}^{\mathrm{BC}}(\mathbf{h}_1, \ldots, \mathbf{h}_K, P) =$$
$$\left\{ (R_1, \ldots, R_K) : R_k \leq \log_2 \left(1 + \frac{P_k ||\mathbf{h}_k||^2}{\sigma^2 + \sum_{j=k+1}^{K} P_j ||\mathbf{h}_k||^2} \right), \sum_{k=1}^{K} P_k \leq P \right\}.$$
$$(3.17)$$

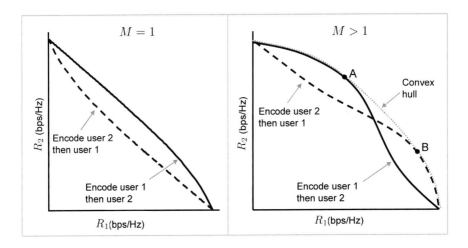

Fig. 3.6 The left subfigure shows the 2-user capacity region for a degraded BC with $M = 1$. The right subfigure shows the 2-user capacity region for a non-degraded BC with $M > 1$. This region is the convex hull of the union of rate regions for the two encoding orders.

3.3.2 Multiple-antenna transmitters $(M > 1)$

In the case of multiple-antenna transmitters, $M > 1$, the users' channels can no longer be absolutely ranked in terms of their channel strengths because the additional degrees of freedom allow the transmitter to spatially steer power between users. For example, consider the case of two single-antenna users with $1 \times M$ MISO channels $\mathbf{h}_1^H$ and $\mathbf{h}_2^H$ such that $||\mathbf{h}_1|| \leq ||\mathbf{h}_2||$. We could weight the codewords u_1 and u_2 for users with respective M-dimensional unit-power beamforming vectors $\mathbf{g}_1$ and $\mathbf{g}_2$ so that $|\mathbf{h}_1^H \mathbf{g}_1| > |\mathbf{h}_2^H \mathbf{g}_2|$, which is opposite to the MISO channel ranking. Since an absolute ranking of the user channels no longer exists, the broadcast channel for $M > 1$ is said to be *non-degraded*. For such channels, superposition coding is no longer capacity-achieving; however, DPC achieves the capacity region of the MIMO BC [80].

Let us consider a $(4,(2,1))$ broadcast channel with a $M = 4$-antenna transmitter serving two single-antenna users. We assume that the channels $\mathbf{h}_1$ and $\mathbf{h}_2$, and beamforming vectors $\mathbf{g}_1$ and $\mathbf{g}_2$, are fixed. If user 1 is encoded first, the transmitter first picks a codeword u_1. The transmit covariance is $\mathbf{Q}_1 := \mathbf{g}_1 \mathbf{g}_1^H \mathbb{E} \left(||u_1||^2 \right)$. With knowledge of u_1, $\mathbf{Q}_1$, and $\mathbf{h}_2$, the transmitter picks codeword u_2 using DPC. The transmit covariance for user 2 is $\mathbf{Q}_2$, and the total transmit power is constrained so that $tr(\mathbf{Q}_1) + tr(\mathbf{Q}_2) \leq P$. As a result of DPC, the rate pairs (R_1, R_2) satisfy

$$R_1 \leq \log_2 \left(\frac{1 + \frac{1}{\sigma^2} \mathbf{h}_1^H (\mathbf{Q}_1 + \mathbf{Q}_2) \mathbf{h}_1}{1 + \frac{1}{\sigma^2} \mathbf{h}_1^H \mathbf{Q}_2 \mathbf{h}_1} \right) \tag{3.18}$$

$$R_2 \leq \log_2 \left(1 + \frac{1}{\sigma^2} \mathbf{h}_2^H \mathbf{Q}_2 \mathbf{h}_2 \right). \tag{3.19}$$

Switching the encoding order, the rate pairs satisfy

$$R_1 \leq \log_2 \left(1 + \frac{1}{\sigma^2} \mathbf{h}_1^H \mathbf{Q}_1 \mathbf{h}_1 \right) \tag{3.20}$$

$$R_2 \leq \log_2 \left(\frac{1 + \frac{1}{\sigma^2} \mathbf{h}_2^H (\mathbf{Q}_1 + \mathbf{Q}_2) \mathbf{h}_2}{1 + \frac{1}{\sigma^2} \mathbf{h}_2^H \mathbf{Q}_1 \mathbf{h}_2} \right). \tag{3.21}$$

The right subfigure of Figure 3.6 shows the rate regions achieved with the two encoding orders, where different vectors on the boundary are achieved by varying the power split between the two users. Unlike the degraded BC case, the regions are neither convex nor concave, and the union of these regions is not convex. The broadcast channel capacity region is the *convex hull* (in other words, the minimal convex set) of the union of these two rate regions.

Rate vectors in the convex hull that lie outside these two rate regions (for example, those vectors on the segment between points A and B in the right subfigure of Figure 3.6) can be achieved using time-sharing.

In general, for multiple transmit antennas $M > 1$ and multiple receive antennas $N > 1$, the broadcast channel capacity region is the convex hull of the union of rate regions corresponding to all possible encoding permutations:

$$\mathcal{C}^{BC}\left(\{\mathbf{H}_k^H\}, P/\sigma^2\right) = \text{Co}\left(\bigcup_\pi \mathbf{R}_\pi\right), \tag{3.22}$$

where Co denotes the convex hull operation and $\mathbf{R}_\pi$ denotes the rate region for a given encoding permutation π. This rate region is given as

$$\mathbf{R}_\pi = \left\{ (R_{\pi(1)}, \ldots, R_{\pi(K)}) : R_{\pi(k)} = \right.$$

$$\left. \log_2 \frac{\det\left(\mathbf{I}_N + \frac{1}{\sigma^2}\mathbf{H}_{\pi(k)}^H \sum_{j=k}^K \mathbf{Q}_{\pi(j)}\mathbf{H}_{\pi(k)}\right)}{\det\left(\mathbf{I}_N + \frac{1}{\sigma^2}\mathbf{H}_{\pi(k)}^H \sum_{j=k+1}^K \mathbf{Q}_{\pi(j)}\mathbf{H}_{\pi(k)}\right)}, \sum_{j=1}^K tr\left(\mathbf{Q}_j\right) \le P \right\}, \tag{3.23}$$

where $\pi(k)$ $(k = 1, \ldots, K)$ denotes the index of the kth encoded user.

For a given permutation π and a given set of transmit covariances, the received signal by user $\pi(k)$ experiences no interference from those users encoded before it $(\pi(1), \ldots, \pi(k-1))$ as a result of DPC. Therefore from (3.2), its virtual received signal is

$$\mathbf{x}_{\pi(k)} = \mathbf{H}_{\pi(k)}^H \mathbf{s} + \sum_{j=k+1}^K \mathbf{H}_{\pi(j)}^H \mathbf{s} + \mathbf{n}_{\pi(k)}, \tag{3.24}$$

where the interference is caused by signals intended for users $\pi(k+1), \ldots, \pi(K)$. From (3.23), user $\pi(k)$ can achieve a rate of

$$R_{\pi(k)} = \log_2 \frac{\det\left(\mathbf{I}_N + \frac{1}{\sigma^2}\mathbf{H}_{\pi(k)}^H \sum_{j=k}^K \mathbf{Q}_{\pi(j)}\mathbf{H}_{\pi(k)}\right)}{\det\left(\mathbf{I}_N + \frac{1}{\sigma^2}\mathbf{H}_{\pi(k)}^H \sum_{j=k+1}^K \mathbf{Q}_{\pi(j)}\mathbf{H}_{\pi(k)}\right)}$$

$$= \log_2 \det\left[\mathbf{I}_N + \left(\sigma^2\mathbf{I}_N + \sum_{j=k+1}^K \mathbf{H}_{\pi(k)}^H \mathbf{Q}_{\pi(j)}\mathbf{H}_{\pi(k)}\right)^{-1} \mathbf{H}_{\pi(k)}^H \mathbf{Q}_{\pi(k)}\mathbf{H}_{\pi(k)}\right]$$

by whitening its virtual received signal (3.24) with respect to its noise and interference covariance and then performing MMSE-SIC with respect to its desired symbol vector $\mathbf{u}_{\pi(k)}$. Figure 3.7 shows a block diagram for achieving the BC capacity region (3.22) using DPC and precoding at the transmitter and MMSE-SIC at each receiver.

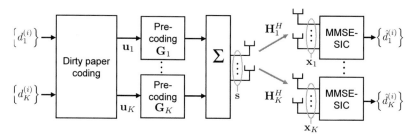

Fig. 3.7 A capacity-achieving strategy for the MIMO BC. The transmitter uses dirty paper encoding to jointly encode the users' data streams. The symbol vector $\mathbf{u}_k$ for user k is precoded using the matrix $\mathbf{G}_k$. At the receiver for user k, the signal is whitened with respect to interference from signals not eliminated from DPC, and MMSE-SIC is performed with respect to the desired symbol vector $\mathbf{u}_k$.

3.4 MAC-BC Duality

In this section, we describe a useful duality [81] between the MIMO MAC capacity region (3.10) and the MIMO BC capacity region (3.22) that provides valuable insights and a tool for evaluating the performance of a MIMO BC channel (described in Section 3.5). For a given $(M, (K, N))$ MIMO BC where the kth user's channel is an $N \times M$ matrix $\mathbf{H}_k^H$, we define a dual $((K, N), M)$ MIMO MAC where the kth user's channel is the $M \times N$ matrix $\mathbf{H}_k$. The following MAC-BC duality characterizes the MIMO BC capacity region in terms of the dual MIMO MAC capacity region:

$$\mathcal{C}^{\mathrm{BC}}\left(\{\mathbf{H}_k^H\}, P/\sigma^2\right) = \bigcup_{P_k \geq 0, \sum_{k=1}^{K} P_k \leq P} \mathcal{C}^{\mathrm{MAC}}\left(\{\mathbf{H}_k\}, \{P_k/\sigma^2\}\right). \quad (3.25)$$

It states that any rate vector in the $(M, (K, N))$ BC capacity region with power constraint P can be obtained on the dual $((K, N), M)$ MAC if the mobiles are allowed to distribute power P among themselves. As an example, Figure 3.8 shows the BC and dual MAC capacity regions for $K = 2$, $M =$

$N = 1$. The BC rate region with $P = 2$ is the union of the dual MAC capacity regions where the power is split between the users.

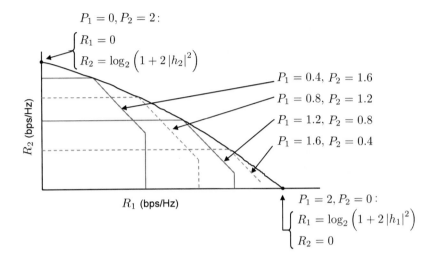

Fig. 3.8 $K = 2$-user BC capacity region, shown as the union of the dual MAC capacity regions with different power splits between the users.

The proof of the duality relies on a transformation between the covariances for the MAC and the BC, where, for any set of MAC covariances $\mathbf{Q}_1^{MAC}, ..., \mathbf{Q}_K^{MAC}$ of size $N \times N$ and any decoding order for the users, there exists a set of BC covariances $\mathbf{Q}_1^{BC}, ..., \mathbf{Q}_K^{BC}$ of size $M \times M$ with the same sum power such that the rates achieved by each of the K users are the same. The converse also holds, so that for any set of BC covariances and any encoding order, there exists a set of MAC covariances with the same sum power such that the user rates are the same.

In particular, the transformation relies on a reversal of the decoding and encoding orders. For example, if users on the MAC are decoded in order starting with user 1 and ending with user K (so that user K sees no interference), then the users on the BC are encoded in reverse order from user K to user 1 (so that user 1 sees no interference). With this decoding order, the rate for user k on the MAC is

$$R_k^{\mathrm{MAC}} = \log_2 \frac{\det\left(\mathbf{I}_M + \frac{1}{\sigma^2}\sum_{j=k}^{K}\mathbf{H}_j\mathbf{Q}_j^{\mathrm{MAC}}\mathbf{H}_j^H\right)}{\det\left(\mathbf{I}_M + \frac{1}{\sigma^2}\sum_{j=k+1}^{K}\mathbf{H}_j\mathbf{Q}_j^{\mathrm{MAC}}\mathbf{H}_j^H\right)}. \tag{3.26}$$

And the rate for user k on the BC is

$$R_k^{\mathrm{BC}} = \log_2 \frac{\det\left(\mathbf{I}_N + \frac{1}{\sigma^2}\mathbf{H}_k^H\sum_{j=1}^{k}\mathbf{Q}_j^{\mathrm{BC}}\mathbf{H}_k\right)}{\det\left(\mathbf{I}_N + \frac{1}{\sigma^2}\mathbf{H}_k^H\sum_{j=1}^{k-1}\mathbf{Q}_j^{\mathrm{BC}}\mathbf{H}_k\right)}. \tag{3.27}$$

Using the duality transformation, the sum powers are the same:

$$\sum_{k=1}^{K}\operatorname{tr}\mathbf{Q}_k^{\mathrm{BC}} = \sum_{k=1}^{K}\operatorname{tr}\mathbf{Q}_k^{\mathrm{MAC}}, \tag{3.28}$$

and user rates are the same: $R_k^{\mathrm{MAC}} = R_k^{\mathrm{BC}}$, $k = 1, 2, ..., K$.

3.5 Scalar performance metrics

Given a K-dimensional capacity region $\mathcal{C}^{\mathrm{MAC}}$ or $\mathcal{C}^{\mathrm{BC}}$, it would be useful to define scalar performance metrics that map the multidimensional region to a single quantity in order to make performance comparisons between different strategies. The sum-rate capacity is one such metric which we defined for the MAC (3.4), and we could define it similarly for the BC capacity region:

$$\max_{\mathbf{R}\in\mathcal{C}^{\mathrm{BC}}}\sum_{k=1}^{K}R_k. \tag{3.29}$$

More generally, we could define a *weighted sum-rate* metric

$$\max_{\mathbf{R}\in\mathcal{C}}\sum_{k=1}^{K}q_k R_k, \tag{3.30}$$

where $\mathcal{C}$ denotes either the MAC or BC capacity region and where the weights are non-negative constants $q_k \geq 0, k = 1, \ldots, K$. We will see in Section 5.4.2 how these weights can be set to achieve quality-of-service requirements in the context of scheduling and resource allocation. For this reason, we refer to the weights $q_1, \ldots, q_K$ as quality-of-service (QoS) weights. A user with a higher QoS weight has higher relative importance in the sense that it contributes more to the weighted sum-rate metric. In practice, we are interested not in the weighted sum rate but in the rate vector which maximizes the weighted sum rate:

$$\mathbf{R}^{\text{Opt}} = \arg\max_{\mathbf{R} \in \mathcal{C}} \sum_{k=1}^{K} q_k R_k. \tag{3.31}$$

Because the capacity regions are convex and the weights are nonnegative, the desired rate vector $\mathbf{R}^{\text{Opt}}$ lies on the boundary of the rate region.

Note that as a consequence of the MAC-BC duality, given P and a set of QoS weights, the BC weighted sum rate is an upper bound on the MAC weighted sum rate for any power partition $P_1, \ldots, P_K$ such that $\sum_{k=1}^{K} P_k = P$:

$$\max_{\mathbf{R} \in \mathcal{C}^{\text{BC}}\left(\{\mathbf{H}_k^H\}, P/\sigma^2\right)} \sum_{k=1}^{K} q_k R_k \geq \max_{\mathbf{R} \in \mathcal{C}^{\text{MAC}}\left(\{\mathbf{H}_k\}, \{P_k/\sigma^2\}\right)} \sum_{k=1}^{K} q_k R_k. \tag{3.32}$$

In the context of a single base station serving multiple users, if the base transmit power is equal to the sum of the user transmit powers, the downlink performance will be at least as good as the uplink performance using the reciprocal channels.

3.5.1 Maximizing the MAC weighted sum rate

Achieving the optimal rate vector that maximizes the weighted sum rate (3.31) is straightforward for the MAC with fixed covariances because, as a result of the polymatroid shape of the rate region, the optimal rate vector is one of the vertices of the sum-rate capacity hyperplane. To lie on this hyperplane, all users transmit with a fixed covariance at full power, and the desired vertex is given by the cancelation order related to the ranking of the QoS weights [78]. Without loss of generality, if $q_1 \leq q_2 \leq \cdots \leq q_K$, then the optimal rate vector $\mathbf{R}^{\text{Opt}}$ which maximizes (3.31) is given by the vertex where the users are decoded in order from 1 to K so that user 1 experiences interference from all other users, and user K experiences no interference. The rate R_k for user k $(k = 1, \ldots, K)$ is given by (3.12).

While a network operator may be interested in determining the rate vector $\mathbf{R}$ that maximizes the weighted sum rate, the actual measured performance in terms of throughput is given by the sum rate $\sum_{k=1}^{K} R_k$. In summing these rates given by (3.12), the denominator for R_k $(k = 1, \ldots, K-1)$ cancels with the numerator for R_{k+1}. The sum rate with fixed covariances is therefore given just by the numerator of R_1:

$$\max_{\mathbf{R} \in \mathcal{C}^{\text{MAC}}(\{\mathbf{H}_k\}, \{\mathbf{Q}_k/\sigma^2\})} \sum_{k=1}^{K} R_k = \log_2 \det \left(\mathbf{I}_M + \frac{1}{\sigma^2} \sum_{j=1}^{K} \mathbf{H}_j \mathbf{Q}_j \mathbf{H}_j^H \right). \quad (3.33)$$

Note that while the individual user rates in (3.12) were determined by the decoding order, the sum rate (3.33) is independent of the ordering because any vertex maximizing the weighted sum rate lies on the sum-rate capacity hyperplane.

We can visualize this concept for the $K = 2$-user MAC rate region shown in Figure 3.9. For fixed q_1 and q_2, contour lines for a given weighted sum rate have slope $-q_1/q_2$. Therefore if $q_2 > q_1$, the optimal rate vector is achieved by decoding user 1 first. On the other hand, if $q_1 > q_2$, the optimal rate vector is achieved by decoding user 2 first. In general, the user with the largest QoS weight is decoded last so that it experiences no interference.

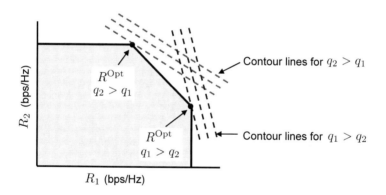

Fig. 3.9 MAC capacity region and contour lines for weighted sum rates given q_1 and q_2. The rate vectors that maximize the weighted sum rate are given by $\mathbf{R}^{\text{Opt}}$.

To maximize the weighted sum rate when the transmit covariances are not specified, the optimal decoding order is still given by the ranking of the QoS weights because this ordering is the best once the covariances are fixed. If $q_1 \leq q_2 \leq \dots \leq q_K$, the maximum weighted sum rate is obtained by optimizing over the transmit covariances, subject to the power constraint for each user:

$$\max_{\mathbf{R} \in \mathcal{C}^{\mathrm{MAC}}(\{\mathbf{H}_k\}, \{P_k/\sigma^2\})} \sum_{k=1}^{K} q_k R_k =$$

$$\max_{\mathrm{tr}\, \mathbf{Q}_k \leq P_k} \sum_{k=1}^{K} q_k \log_2 \frac{\det\left(\mathbf{I}_M + \frac{1}{\sigma^2} \sum_{j=k}^{K} \mathbf{H}_j \mathbf{Q}_j \mathbf{H}_j^H\right)}{\det\left(\mathbf{I}_M + \frac{1}{\sigma^2} \sum_{j=k+1}^{K} \mathbf{H}_j \mathbf{Q}_j \mathbf{H}_j^H\right)}. \quad (3.34)$$

Iterative algorithms to determine the optimal covariances can be found in [82] for sum rate maximization, and in [83] for weighted sum rate maximization.

Because the weighted sum rate maximization problem (3.34) is a convex optimization, conventional numerical optimization techniques can also be used [84]. For example, the following gradient-based optimization simplifies the problem by reducing the multidimensional optimization in each step to a single-dimensional optimization [85]. We assume without loss of generality that the QoS weights are ordered: $q_1 \leq q_2 \leq \ldots \leq q_K$. We define the weighted sum rate objective function as a function of the covariances $\mathbf{Q}_k^{(n)}$, $k = 1, \ldots, K$, during iteration n:

$$f(\mathbf{Q}_1^{(n)}, \ldots, \mathbf{Q}_K^{(n)}) := \sum_{k=1}^{K} q_k \log_2 \frac{\det\left(\mathbf{I}_M + \frac{1}{\sigma^2} \sum_{j=k}^{K} \mathbf{H}_j^H \mathbf{Q}_j^{(n)} \mathbf{H}_j\right)}{\det\left(\mathbf{I}_M + \frac{1}{\sigma^2} \sum_{j=k+1}^{K} \mathbf{H}_j^H \mathbf{Q}_j^{(n)} \mathbf{H}_j\right)} \quad (3.35)$$

$$= q_1 \log_2 \det\left(\mathbf{I}_M + \frac{1}{\sigma^2} \sum_{j=1}^{K} \mathbf{H}_j^H \mathbf{Q}_j^{(n)} \mathbf{H}_j\right) +$$

$$\sum_{k=2}^{K} (q_k - q_{k-1}) \log_2 \det\left(\mathbf{I}_M + \frac{1}{\sigma^2} \sum_{j=k}^{K} \mathbf{H}_j^H \mathbf{Q}_j^{(n)} \mathbf{H}_j\right).$$

We also define the gradient of the objective function with respect to the ith covariance $\mathbf{Q}_i$, $i = 1, \ldots, K$:

$$\nabla f_i(\mathbf{Q}_1^{(n)}, \ldots, \mathbf{Q}_K^{(n)}) := q_1 \left[\mathbf{H}_i^H \left(\mathbf{I}_M + \frac{1}{\sigma^2} \sum_{j=1}^{K} \mathbf{H}_j^H \mathbf{Q}_j^{(n)} \mathbf{H}_j\right)^{-1} \mathbf{H}_i\right] +$$

$$\sum_{k=2}^{i} (q_k - q_{k-1}) \left[\mathbf{H}_i \left(\mathbf{I}_M + \frac{1}{\sigma^2} \sum_{j=k}^{K} \mathbf{H}_j^H \mathbf{Q}_j^{(n)} \mathbf{H}_j\right)^{-1} \mathbf{H}_i^H\right]. \quad (3.36)$$

The following algorithm proceeds iteratively until the covariance matrices converge to the optimal covariance matrices.

Algorithm for computing optimal MAC covariances

Let $n := 1$

Initialize $\mathbf{Q}_k^{(n)} := \mathbf{0}$, $k = 1, \ldots, K$

while not converged

    for $k = 1$ **to** K

        a. Let λ_k^2 be the dominant eigenvalue of

$$\nabla f_k \left(\mathbf{Q}_1^{(n+1)}, \ldots, \mathbf{Q}_{k-1}^{(n+1)}, \mathbf{Q}_k^{(n)}, \mathbf{Q}_{k+1}^{(n)}, \ldots, \mathbf{Q}_K^{(n)} \right)$$

        and let $\mathbf{v}_k$ be the corresponding eigenvector

        b. Find t^* which is the solution to the one-dimensional
        optimization:

$$t^* = \arg\max_{t \in [0,1]} f \left(\mathbf{Q}_1^{(n+1)}, \ldots, \mathbf{Q}_{k-1}^{(n+1)}, \right.$$
$$\left. t\mathbf{Q}_k^{(n)} + (1-t)P_k \mathbf{v}_k \mathbf{v}_k^H, \mathbf{Q}_{k+1}^{(n)}, \ldots, \mathbf{Q}_K^{(n)} \right)$$

        c. Update the kth user's covariance:

$$\mathbf{Q}_k^{(n+1)} := t^* \mathbf{Q}_k^{(n)} + (1-t^*)P_k \mathbf{v}_k \mathbf{v}_k^H$$

    end

    $n := n + 1$

end

3.5.2 Maximizing the BC weighted sum rate

For the MIMO MAC, the weighted sum rate objective is convex with respect to the user covariances. In contrast, for the MIMO BC, this metric is in general not convex with respect to the transmit covariances, making the direct computation of the maximum weighted sum rate difficult. However, as a result of the MAC-BC duality in Section 3.4, the MIMO BC weighted sum rate can be characterized as a convex function in terms of the dual MAC transmit covariances. Using the duality, the maximum weighted sum rate for the MIMO BC can be written as

$$\max_{\mathbf{R} \in \mathcal{C}^{BC}(\{\mathbf{H}_k\}, P/\sigma^2)} \sum_{k=1}^K q_k R_k = \max_{\mathbf{R} \in \bigcup_{P_k, \sum P_k \leq P} \mathcal{C}^{MAC}(\{\mathbf{H}_k\}, \{P_k/\sigma^2\})} \sum_{k=1}^K q_k R_k$$

$$= \max_{P_k, \sum P_k \leq P} \max_{\operatorname{tr} \mathbf{Q}_k \leq P_k} \sum_{k=1}^{K} q_k \log_2 \frac{\det \left(\mathbf{I}_M + \frac{1}{\sigma^2} \sum_{j=k}^{K} \mathbf{H}_j \mathbf{Q}_j \mathbf{H}_j^H \right)}{\det \left(\mathbf{I}_M + \frac{1}{\sigma^2} \sum_{j=k+1}^{K} \mathbf{H}_j \mathbf{Q}_j \mathbf{H}_j^H \right)}. \quad (3.37)$$

A variation of the technique for maximizing the MIMO MAC weighted sum rate (Section 3.5.1) can be used for the MIMO BC [85]. We use the same objective function (3.35) and gradient (3.36), where $\mathbf{H}_k^H$ is the $M \times N$ dual MAC channel for the kth user. The covariances $\mathbf{Q}_k^{(n)}$ are again size $N \times N$. Compared to the algorithm for the MAC in Section 3.5.1, the following iterative algorithm for the dual MAC allows for power sharing among the users.

Algorithm for computing optimal dual MAC covariances

Let $n := 1$

Initialize $\mathbf{Q}_k^{(n)} := \mathbf{0}$, $k = 1, \ldots, K$

while not converged

    a. For each user $k = 1, \ldots, K$, let λ_k^2 be the dominant eigenvalue of

$$\nabla f_k \left(\mathbf{Q}_1^{(n)}, \ldots, \mathbf{Q}_{k-1}^{(n)}, \mathbf{Q}_k^{(n)}, \mathbf{Q}_{k+1}^{(n)}, \ldots, \mathbf{Q}_K^{(n)} \right)$$

    and let $\mathbf{v}_k$ be the corresponding eigenvector

    b. Let $i^* = \arg\max(\lambda_1^2, \ldots, \lambda_K^2)$

    c. Find t^* which is the solution to the one-dimensional optimization:

$$t^* = \arg\max_{t \in [0,1]} f \left(t\mathbf{Q}_1^{(n)}, \ldots, t\mathbf{Q}_{i^*-1}^{(n)}, \right.$$
$$\left. t\mathbf{Q}_{i^*}^{(n)} + (1-t)P_k \mathbf{v}_{i^*} \mathbf{v}_{i^*}^H, t\mathbf{Q}_{i^*+1}^{(n)}, \ldots, t\mathbf{Q}_K^{(n)} \right)$$

    d. Update the covariances:

$$\mathbf{Q}_{i^*}^{(n+1)} := t^* \mathbf{Q}_{i^*}^{(n)} + (1-t^*)P \mathbf{v}_{i^*} \mathbf{v}_{i^*}^H$$
$$\mathbf{Q}_j^{(n+1)} := t^* \mathbf{Q}_j^{(n)}, j \neq i^*$$

    e. $n := n + 1$

end

After the algorithm converges, the BC covariances (of size $M \times M$) are computed from the dual MAC covariances (of size $N \times N$) via the duality transformation [81]. Recall that for a given decoding order of the dual MAC users, the corresponding BC user rates are achieved by reversing the order for the downlink encoding. Therefore, given the ordered QoS weights $q_1 \leq$

$q_2 \leq \ldots \leq q_K$, the MAC weighted sum rate is maximized by decoding the users in order starting with user 1 and ending with user K, and the BC weighted sum rate is maximized by encoding the users in order starting with user K and ending with user 1. As highlighted in Figure 3.10, the user with the highest QoS weight (user K) experiences no interference on the uplink but experiences interference from all other users on the downlink.

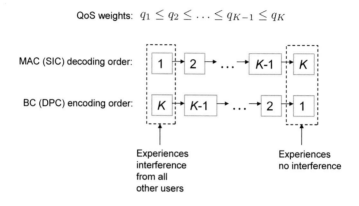

Fig. 3.10 To maximize the weighted sum rate for the MAC, users are decoded with successive interference cancelation (SIC) in the order of *ascending* QoS weights. For the BC, users are encoded with dirty paper coding (DPC) in the order of *descending* QoS weights.

3.6 Sum-rate performance

In this section, we study the sum-rate capacity performance of the MIMO MAC and BC. Our observations will provide motivation for designing suboptimal transceiver techniques (Chapter 5) and provide some insights into the cellular system performance when users have different geometries (Chapter 6). We use the $((K, N), M)$ MAC signal model (3.1) and the $(M, (K, N))$ BC signal model (3.2). The BC has transmit power P, and each user on the MAC has transmit power P/K, so that the total power is the same for the two systems. We use the following notation to denote the sum-rate capacity of the MAC for channels $\mathbf{H}_1, \ldots, \mathbf{H}_K$ and of the BC for channels $\mathbf{H}_1^H, \ldots, \mathbf{H}_K^H$:

$$C^{\mathrm{MAC}}\left(\{\mathbf{H}_k\}, P/\sigma^2\right) := \max_{\mathbf{R}\in\mathcal{C}^{\mathrm{MAC}}(\{\mathbf{H}_k\},\{P/(K\sigma^2)\})} \sum_{k=1}^{K} R_k \qquad (3.38)$$

$$C^{\mathrm{BC}}\left(\{\mathbf{H}_k^H\}, P/\sigma^2\right) := \max_{\mathbf{R}\in\mathcal{C}^{\mathrm{BC}}(\{\mathbf{H}_k^H\}, P/\sigma^2)} \sum_{k=1}^{K} R_k. \qquad (3.39)$$

As an alternative way to communicate over the MIMO BC, we consider time-division multiple access (TDMA) transmission where users are served one at a time. With knowledge of CSI at the transmitter, the maximum sum rate for TDMA is achieved by serving the user with the highest CL-MIMO capacity:

$$C^{\mathrm{TDMA}}\left(\{\mathbf{H}_k^H\}, P/\sigma^2\right) := \max_k \max_{\mathrm{tr}\,\mathbf{Q}=P} \log\det\left(\mathbf{I} + \frac{1}{\sigma^2}\mathbf{H}_k\mathbf{Q}\mathbf{H}_k^H\right). \qquad (3.40)$$

This strategy is also referred to as time-sharing [86].

As another performance benchmark, we consider the open-loop and closed-loop capacity of a single-user channel $\mathbf{H} \in \mathbb{C}^{M\times KN}$, which is the concatenation of the multiuser channels:

$$\mathbf{H} := [\mathbf{H}_1, \ldots, \mathbf{H}_K]. \qquad (3.41)$$

With power constraint P, we define the capacities of the corresponding single-user channels as:

$$C^{\mathrm{OL}}(\mathbf{H}, P/\sigma^2) := \log_2\det\left(\mathbf{I}_M + \frac{P}{\sigma^2 KN}\mathbf{H}\mathbf{H}^H\right) \qquad (3.42)$$

$$C^{\mathrm{CL}}(\mathbf{H}, P/\sigma^2) := \max_{\mathbf{Q}>0,\mathrm{tr}\,\mathbf{Q}=P} \log_2\det\left(\mathbf{I}_M + \frac{1}{\sigma^2}\mathbf{H}\mathbf{Q}\mathbf{H}^H\right). \qquad (3.43)$$

Given the sum-rate metrics defined for a given channel realization, we can define the respective average sum rates where the expectation is taken over random channel realizations:

$$\bar{C}^{\mathrm{MAC}}((K,N), M, P/\sigma^2) := \mathbb{E}_{\{\mathbf{H}_k\}}\left[C^{\mathrm{MAC}}\left(\{\mathbf{H}_k\}, P/\sigma^2\right)\right] \qquad (3.44)$$

$$\bar{C}^{\mathrm{BC}}(M, (K,N), P/\sigma^2) := \mathbb{E}_{\{\mathbf{H}_k^H\}}\left[C^{\mathrm{BC}}\left(\{\mathbf{H}_k^H\}, P/\sigma^2\right)\right] \qquad (3.45)$$

$$\bar{C}^{\mathrm{TDMA}}(M, (K,N), P/\sigma^2) := \mathbb{E}_{\{\mathbf{H}_k^H\}}\left[C^{\mathrm{TDMA}}\left(\{\mathbf{H}_k^H\}, P/\sigma^2\right)\right]. \qquad (3.46)$$

In the numerical results that follow, we assume the channels are i.i.d. Rayleigh.

3.6.1 MU-MIMO sum-rate capacity and SU-MIMO capacity

For a given set of channels $\mathbf{H}_1, \ldots, \mathbf{H}_K$, we can establish a useful relationship among the MIMO MAC sum-rate capacity (3.38), MIMO BC sum-rate capacity (3.39), and the capacity of the corresponding SU-MIMO channels (3.43) and (3.42). Let us first define a $((KN, 1), M)$ MAC for KN single-antenna users where the SIMO channels of each user are given by the columns of $[\mathbf{H}_1, \ldots, \mathbf{H}_K]$ and where the transmit power of each user is $P/(KN)$. We let $C^{\mathrm{MAC}}(\mathbf{h}_{11}, \ldots, \mathbf{h}_{1N}, \ldots, \mathbf{h}_{K1}, \ldots, \mathbf{h}_{KN}, P/\sigma^2)$ be the sum-rate capacity of this MAC. Due to the equivalence between the open-loop MIMO capacity and the MAC with equal-power transmitters (Section 3.2.1), we can use (3.42) to write

$$C^{\mathrm{OL}}(\mathbf{H}, P/\sigma^2) = C^{\mathrm{MAC}}(\mathbf{h}_{11}, \ldots, \mathbf{h}_{1N}, \ldots, \mathbf{h}_{K1}, \ldots, \mathbf{h}_{KN}, P/\sigma^2). \quad (3.47)$$

By grouping together K sets of N SIMO channels so that the MIMO channel for user k is $\mathbf{H}_k$, we get a new MAC that consists of K users each with N antennas and power P/K. This $((K, N), M)$ MAC could achieve a higher sum-rate capacity because the transmissions across the N antennas for each user are now coordinated:

$$C^{\mathrm{MAC}}(\mathbf{h}_{11}, \ldots, \mathbf{h}_{1N}, \ldots, \mathbf{h}_{K1}, \ldots, \mathbf{h}_{KN}, P/\sigma^2) \leq C^{\mathrm{MAC}}\left(\{\mathbf{H}_k\}, P/\sigma^2\right). \quad (3.48)$$

From the MAC-BC duality (3.32), the BC sum-rate capacity is greater than or equal to the dual MAC sum-rate capacity for a given power partitioning. Therefore

$$C^{\mathrm{MAC}}\left(\{\mathbf{H}_k\}, P/\sigma^2\right) \leq C^{\mathrm{BC}}\left(\{\mathbf{H}_k^H\}, P/\sigma^2\right). \quad (3.49)$$

Given a $(M, (K, N))$ BC with channels $\mathbf{H}_1^H, \ldots, \mathbf{H}_K^H$, an upper bound on the sum rate can be obtained by coordinating the detection across the K users. This upper bound can be characterized by the closed-loop MIMO capacity of the channel $\mathbf{H}^H$ defined in (3.41):

$$C^{\mathrm{BC}}\left(\{\mathbf{H}_k^H\}, P/\sigma^2\right) \leq C^{\mathrm{CL}}(\mathbf{H}^H, P/\sigma^2). \quad (3.50)$$

Because $C^{\mathrm{CL}}(\mathbf{H}^H, P/\sigma^2) = C^{\mathrm{CL}}(\mathbf{H}, P/\sigma^2)$, we can combine the results from with (3.47),(3.48),(3.49), and (3.50) to obtain a chain of inequalities:

$$C^{\mathrm{OL}}(\mathbf{H}, P/\sigma^2) \leq C^{\mathrm{MAC}}\left(\{\mathbf{H}_k\}, P/\sigma^2\right)$$

$$\leq C^{\mathrm{BC}}\left(\left\{\mathbf{H}_k^H\right\}, P/\sigma^2\right)$$
$$\leq C^{\mathrm{CL}}(\mathbf{H}, P/\sigma^2). \tag{3.51}$$

For the special case of $N = 1$, the OL-MIMO capacity and MAC sum-rate capacity are equivalent, resulting in the following:

$$C^{\mathrm{OL}}(\mathbf{H}, P/\sigma^2) = C^{\mathrm{MAC}}\left(\left\{\mathbf{h}_k\right\}, P/\sigma^2\right)$$
$$\leq C^{\mathrm{BC}}\left(\left\{\mathbf{h}_k^H\right\}, P/\sigma^2\right)$$
$$\leq C^{\mathrm{CL}}(\mathbf{H}, P/\sigma^2). \tag{3.52}$$

Recall that at asymptotically high SNR, the open- and closed-loop capacities for a SU-MIMO channel are equivalent because CSIT does not provide any performance benefit (Section 2.2.2.2). It follows from (3.51) that at asymptotically high SNR, the MIMO MAC and MIMO BC sum-rate capacities are also equivalent. Therefore for a fixed (and high) SNR P/σ^2, the sum-rate capacity does not depend on the direction of the transmission. Figure 1.9 shows the convergence of the CL-MIMO, MAC, and BC capacities at high SNR. On the other hand, at low SNRs, CSIT for the SU-MIMO channel provides an advantage, and there can be a significant difference in the MAC and BC sum-rate capacity performance, as seen in Figure 1.10.

3.6.2 Sum rate versus SNR

By studying the sum-rate capacity performance of MU-MIMO channels more carefully for low and high SNRs, we can gain insights into transmit strategies for, respectively, low and high user geometries in cellular networks.

3.6.2.1 MAC sum rate at low SNR

We first consider the sum-rate capacity of the equal-power MAC (3.38) for the case of asymptotically low SNR. Following the derivation for the open-loop MIMO capacity for low SNR in Section 2.2.2.1, we have from (3.9) and (3.38):

$$\lim_{P/\sigma^2 \to 0} C^{\mathrm{MAC}}\left(\left\{\mathbf{H}_k\right\}, P/\sigma^2\right) \approx \max_{\mathrm{tr}\,\mathbf{Q}_k \leq P/(\sigma^2 K)} \mathrm{tr}\left(\sum_{k=1}^{K} \mathbf{H}_k \mathbf{Q}_k \mathbf{H}_k^H\right) \log_2 \mathrm{e}$$

$$= \sum_{k=1}^{K} \max_{\mathrm{tr}\,\mathbf{Q}_k \leq P/(\sigma^2 K)} \mathrm{tr}\left(\mathbf{H}_k \mathbf{Q}_k \mathbf{H}_k^H\right) \log_2 \mathrm{e}$$

$$= \sum_{k=1}^{K} \frac{P}{\sigma^2 K} \lambda_{\max}^2\left(\mathbf{H}_k\right) \log_2 \mathrm{e}, \qquad (3.53)$$

where $\lambda_{\max}^2\left(\mathbf{H}_k\right)$ is the maximum eigenvalue of $\mathbf{H}_k \mathbf{H}_k^H$. As the SNR decreases, the additive noise dominates the interference, and it becomes optimal for each user to maximize its own rate without regard to its impact on the others [87]. It does so by transmitting with full power P/K on its dominant eigenmode (3.53). If the channels of the users are statistically equivalent, the average sum-rate capacity is K times the average rate of any user, and it follows that

$$\lim_{P/\sigma^2 \to 0} \frac{\bar{C}^{\mathrm{MAC}}((K,N), M, P/\sigma^2)}{P/\sigma^2} = \mathbb{E}_{\mathbf{H}_k} \lambda_{\max}^2\left(\mathbf{H}_k\right) \log_2 e, \qquad (3.54)$$

where $\mathbb{E}_{\mathbf{H}_k} \lambda_{\max}^2\left(\mathbf{H}_k\right)$ is the expected value of the spectral radius taken over realizations of $\mathbf{H}_k \in \mathbb{C}^{M \times N}$. The expected value can be computed analytically for certain channel distributions including the i.i.d. Rayleigh distribution [88].

3.6.2.2 BC sum rate at low SNR

Recall that for the SU-MIMO channel at asymptotically low SNR, it becomes optimal to transmit a single stream over the best eigenmode if CSIT is known. It follows intuitively that for the BC at low SNR, multiplexing data for multiple users does not provide any advantage over single-user transmission. Indeed, at asymptotically low SNR, the BC capacity region converges to the TDMA achievable rate region [89]. In this regime, the BC sum-rate capacity and maximum TDMA sum rate are equivalent [90] and are achieved by transmitting power P over the best eigenmode among the K users:

$$P/\sigma^2 \max_k \left[\lambda_{\max}^2\left(\mathbf{H}_k^H\right)\right] \log_2 \mathrm{e}. \qquad (3.55)$$

In the limit of very low SNR, the ratio of the average sum rate to the SNR can be expressed as

$$\lim_{P/\sigma^2 \to 0} \frac{\bar{C}^{\mathrm{BC}}(M, (K,N), P/\sigma^2)}{P/\sigma^2} = \lim_{P/\sigma^2 \to 0} \frac{\bar{C}^{\mathrm{TDMA}}(M, (K,N), P/\sigma^2)}{P/\sigma^2}$$

$$= \mathbb{E}_{\{\mathbf{H}_k^H\}} \max_k \left[\lambda_{\max}^2\left(\mathbf{H}_k^H\right)\right] \log_2 e, \qquad (3.56)$$

where the expectation is taken with respect to the channels $\mathbf{H}_1^H, \ldots, \mathbf{H}_K^H$. Because the BC strategy transmits to the best among K users, the average sum-rate capacity (3.56) improves as K increases due to multiuser diversity. In contrast, average MAC sum-rate capacity (3.54) is independent of K (if each user has power P/K).

3.6.2.3 MAC and BC sum rates at high SNR

Shifting to the high SNR regime, we recall that the multiplexing gain for a (M, N) SU-MIMO channel is $\min(M, N)$, whether or not CSIT is available (Section 2.2.2.2). From the chain of inequalities (3.51), it follows that

$$\lim_{P/\sigma^2 \to \infty} \frac{\bar{C}^{\mathrm{MAC}}((K, N), M, P/\sigma^2)}{\log_2(P/\sigma^2)} = \lim_{P/\sigma^2 \to \infty} \frac{\bar{C}^{\mathrm{BC}}(M, (K, N), P/\sigma^2)}{\log_2(P/\sigma^2)}$$
$$= \min(KN, M). \qquad (3.57)$$

Hence the sum-rate capacity of the BC and MAC scales with the minimum of the *total* number of transmit or receive antennas. If $K \geq M$, then increasing the number of terminal antennas N provides combining gain but does not increase the multiplexing order.

Because TDMA serves only a single user, the multiplexing gain is equivalent to a SU-MIMO channel's:

$$\lim_{P/\sigma^2 \to \infty} \frac{\bar{C}^{\mathrm{TDMA}}(M, (K, N), P)}{\log_2(P/\sigma^2)} = \min(M, N). \qquad (3.58)$$

If the BC transmitter does not have CSIT, it may not be able to effectively multiplex signals for multiple users. In particular, for i.i.d. Rayleigh channels, the BC without CSIT achieves the same multiplexing gain of TDMA (3.58) [91], indicating the importance of CSIT for the BC. For the MAC, if CSIT is not available, each user can transmit isotropically to achieve a maximum sum rate of $C^{\mathrm{OL}}(KN, M, P/\sigma^2)$. Therefore the full multiplexing gain $\min(KN, M)$ can be achieved for the MAC without CSIT.

3.6.2.4 Numerical results

We first consider the performance of a BC and dual MAC for $K = 2$ single-antenna users with fixed channel realizations $\mathbf{h}_1$ and $\mathbf{h}_2$. The total power P is the same for the two cases, and it is split evenly for the MAC users. Figure 3.11 shows the rate regions for high and low SNR and for two channel realizations — one where the two channels are nearly orthogonal ($\left|\mathbf{h}_1^H \mathbf{h}_2\right| \ll \|\mathbf{h}_1\|^2 \approx \|\mathbf{h}_1\|^2$) and another where the channels are nearly colinear ($\left|\mathbf{h}_1^H \mathbf{h}_2\right| \approx \|\mathbf{h}_1\|^2 \approx \|\mathbf{h}_1\|^2$).

- As a result of the MAC-BC duality, the MAC capacity region is contained within the BC capacity region in all cases.
- At high SNR for orthogonal channels, interference for the MAC is minimal and each user can achieve its maximum rate simultaneously without interference cancellation: $\log_2(1 + P \|\mathbf{h}_1\|^2) \approx \log_2(1 + P \|\mathbf{h}_1\|^2)$. As a result, the MAC capacity region has a rectangular shape. For the BC, user k ($k = 1, 2$) can achieve rate $\log_2(1 + P \|\mathbf{h}_k\|^2)$ bps/Hz if served alone. By serving both users simultaneously and splitting the power evenly, each user can achieve $\log_2(1 + P/2 \|\mathbf{h}_k\|^2)$, which is only 1 bps/Hz less than the time-multiplexed rate.
- At high SNR for nearly co-linear channels, the interference for the MAC is significant, and interference cancellation provides minimal benefit. The BC capacity region is similar to the TDMA rate region, so superposition coding provides little benefit over time sharing.
- At high SNR regardless of the channel realizations, the MAC and BC sum-rate capacities are similar because the bounding OL and CL MIMO capacities (3.51) are similar.
- At low SNR, the MAC capacity region would have a rectangular shape regardless of the channels because the interference is dominated by the noise power. The BC capacity region is nearly flat and is very similar to the TDMA rate region. The BC has a higher sum-rate capacity, and it is achieved by transmitting with power P to user 1.

Figure 3.12 shows the average sum rate versus SNR for the MAC, BC, and TDMA channels for $K = M$ users, $M = 1$ and 4, and $N = 4$ user antennas. Figure 3.13 shows the same but with $N = 1$. (These two figures can be compared with the corresponding curves for SU-MIMO in Figures 2.5 and 2.6.)

- For $M = 1$, the MAC, BC and TDMA channels are equivalent regardless of N. At high SNR, they all achieve a multiplexing gain of M.
- For $M = 4$ at low SNR, the BC and TDMA sum rates are equivalent because the rate regions are equivalent. At high SNR, the BC and MAC sum rates are equivalent.

- For $M = 4$ and $N = 4$, TDMA achieves the same multiplexing gain as the BC and MAC. The slopes are the same, but there is an offset in the rate as a result of the single-user restriction for TDMA.
- For $M = 4$ and $N = 1$, TDMA cannot achieve multiplexing gain, and its slope is the same as the single-antenna $M = 1$ channels.
- For $M = 4$ at high SNR, additional antennas at the users do not improve the multiplexing gain because it is limited by M: $\min(M, K) = \min(M, 4K) = 4$. However, additional antennas provide higher rates through combining gain. A 6 dB combining gain can be achieved on average by each user, resulting in about a 2 bps/Hz rate improvement. At 20 dB SNR and higher, the net improvement in sum rate in using $N = 4$ antennas per user with $K = 4$ users is indeed about 8 bps/Hz.

Figure 3.14 shows the ratio of MU-MIMO average sum rates (3.44) and (3.45) versus the SU-MIMO average rate, for $M = K = 4$ and $N = 1$ and 4:

$$\frac{\bar{C}^{\mathrm{MAC}}((4, N), 4, P/\sigma^2)}{\bar{C}^{\mathrm{OL}}(1, N, P/\sigma^2)} \quad \text{and} \quad \frac{\bar{C}^{\mathrm{BC}}(4, (4, N), P/\sigma^2)}{\bar{C}^{\mathrm{OL}}(1, N, P/\sigma^2)}. \tag{3.59}$$

(For $N = 1$, comparisons can be made with the SU-MIMO capacity gain in Figure 2.7.) As we will see in Chapter 6, this ratio provides insight into the system tradeoffs between a single MU-MIMO link with $M = 4$ antennas at the base and 4 isolated SU-SIMO links (each with the same power as the MU-MIMO link) with $M = 1$ antenna at the base. The former is superior if the ratio is greater than 4.

- At low SNR, the MAC achieves combining gain compared to the SU channel, and the BC achieves both combining and multiuser diversity gain. This diversity advantage can be significant for very low SNR. The marginal gains of MU-MIMO compared to SU-SIMO are diminished if the mobile has multiple antennas.
- At high SNR, MU-MIMO achieves a multiplexing gain of 4 whereas SU-MIMO achieves a multiplexing gain of 1. Therefore at high SNR, the ratio approaches 4.

Figure 3.15 shows the ratio of the average sum rate between BC (3.45) and TDMA (3.46) for $M = K = 4$ and for $N = 1$ and 4:

$$\frac{\bar{C}^{\mathrm{BC}}(4, (4, N), P/\sigma^2)}{\bar{C}^{\mathrm{TDMA}}(4, (4, N), P/\sigma^2)}. \tag{3.60}$$

- At low SNR, the BC and TDMA sum-rate capacities are equivalent, and the ratio approaches 1 regardless of N.
- At high SNR, the multiplexing gain of the BC and TDMA are given respectively by (3.57) and (3.58), and the ratio is

$$\frac{\min(KN, M)}{\min(M, N)}. \tag{3.61}$$

- For $N = 4$, both TDMA and BC can achieve a multiplexing gain of 4 so the ratio approaches 1. For $N = 1$, TDMA cannot achieve multiplexing gain so the ratio approaches 4. Therefore the relative gains of the capacity-achieving BC techniques versus simple TDMA are diminished at high SNR as the number of user antennas increases.

If we apply these observations to a cellular system, one general conclusion is that if users are located far from the base transmitter (and therefore have low geometries), spatial multiplexing of the users provides minimal gains over simple TDMA transmission. However, if the users are close to the transmitter (and therefore have high geometries), spatial multiplexing offers a significant advantage.

3.6.3 Sum rate versus the number of base antennas M

We consider the sum-rate capacity of the $(M, (K, 1))$ BC and the $((K, 1), M)$ equal-power MAC in the limit of a large number of users and antennas. Because of the equivalence between the (K, M) OL SU-MIMO channel and the $((K, 1), M)$ MAC, in the limit of large K and M with $K/M \to \alpha$, the MAC sum-rate capacity converges almost surely to

$$\lim_{\substack{M,K\to\infty \\ K/M\to\alpha}} \frac{C^{\mathrm{MAC}}\left(\{\mathbf{h}_k\}, P/\sigma^2\right)}{M} = C^{\mathrm{Asym}}(\alpha, P/\sigma^2), \tag{3.62}$$

where the constant on the right side is defined in (2.49). We have the equivalent result for the BC [92]

$$\lim_{\substack{M,K\to\infty \\ K/M\to\alpha}} \frac{C^{\mathrm{BC}}\left(\{\mathbf{h}_k^H\}, P/\sigma^2\right)}{M} = C^{\mathrm{Asym}}(\alpha, P/\sigma^2) \tag{3.63}$$

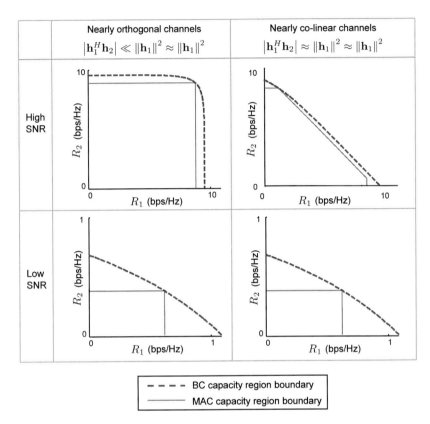

Fig. 3.11 MAC and BC capacity regions for $K = 2$ single-antenna users with fixed channels and with an equal power split for the MAC. At high SNRs, the sum rates of the MAC and BC are equivalent, and they are significantly higher than the TDMA sum rate. At low SNRs, the BC and TDMA sum rates are equivalent, and are slightly higher than the MAC sum rate.

implying that equal distribution of power among users on the dual MAC is asymptotically optimal with respect to the sum-rate capacity scaling. For the $(M, (K, 1))$ TDMA channel, spatial multiplexing cannot be achieved because each user has only a single antenna. Therefore:

$$\lim_{\substack{M,K \to \infty \\ K/M \to \alpha}} \frac{C^{\text{TDMA}}\left(\left\{\mathbf{h}_k^H\right\}, P/\sigma^2\right)}{M} = 0. \qquad (3.64)$$

Figure 3.16 shows the average sum rate of the BC, MAC and TDMA channels versus M for $K = M$, $N = 1$, and i.i.d. Rayleigh channels. For low

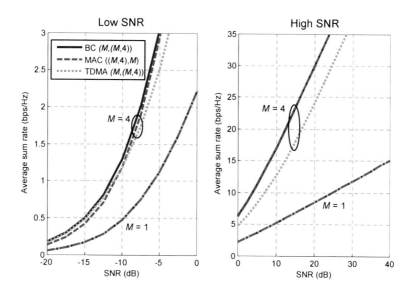

Fig. 3.12 Average sum rate versus SNR for $M = K = 1$ or 4 and for $N = 4$ antennas per user.

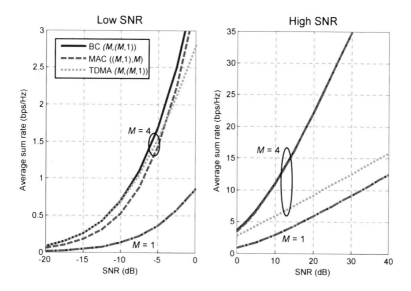

Fig. 3.13 Average sum rate versus SNR for $M = K = 1$ or 4 and for $N = 1$ antennas per user.

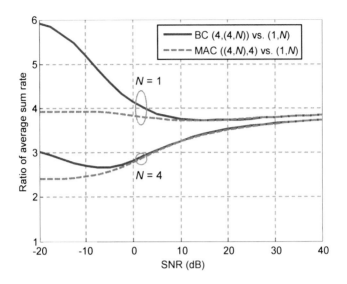

Fig. 3.14 Ratio of MU-MIMO average sum rates versus SU-SIMO average rate as a function of SNR, for $M = K = 4$.

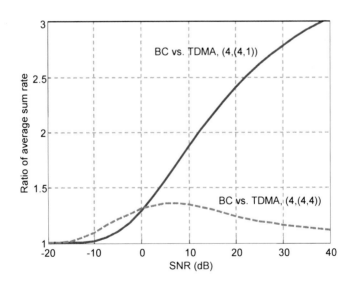

Fig. 3.15 Ratio of BC and TDMA average sum rates as a function of SNR, for $M = K = 4$. At low SNR, the ratio approaches 1, and at high SNR, it approaches $\min(M, KN)/\min(M, N)$.

SNR, the slope of the sum-rate capacity curves for the BC and MAC approach $C^{\text{Asym}}(1, P/\sigma^2)$ as M increases, but there is a gap between the two. TDMA is near optimal for small M. As the SNR increases, the BC-MAC performance gap vanishes, and the multiplexing advantage over TDMA is apparent for small M.

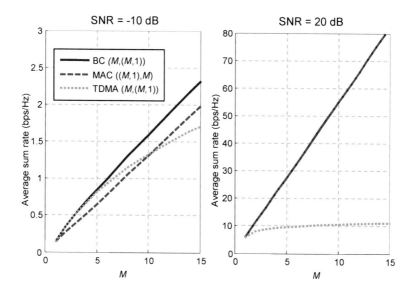

Fig. 3.16 Average sum rate versus M, for $K = M$ and $N = 1$. As M increases asymptotically, the sum-rate slope for both BC and MAC approach $C^*(P)$. The slope for TDMA approaches 0.

3.6.4 Sum rate versus the number of users K

In cellular networks, the number of users K vying for service could potentially be much larger than the number of base antennas M. If we assume $M \geq N$, then in the limit of large K under the assumption of block-fading Rayleigh channels, we have from [86] that

$$\lim_{K \to \infty} \frac{C^{\text{BC}}\left(\{\mathbf{H}_k^H\}, P/\sigma^2\right)}{\log \log KN} = M, \tag{3.65}$$

where the convergence is almost surely. This implies that the multiplexing advantage due to the transmit antennas M dominates over the multiuser diversity achieved with the KN degrees of freedom at the receiver.

The sum-rate scaling in (3.65) can be achieved by treating each N-antenna user as N independent single-antenna users, choosing M random orthonormal beams for transmission, and in each of these beams transmitting to the user antenna with the highest SINR [93]. When M and P are fixed and K is large, the maximum SINR in each beam can be shown to be $P \log(KN) + O(\log \log(KN))$. This leads to a sum rate of $M \log(P \log(KN)) + MO(\log \log \log(KN))$. The proof that this is the fastest scaling possible can be found in [86].

It was shown in [94] that for $N = 1$ and correlated base antennas, there is degradation in the sum rate, but the sum-rate scaling is the same as for the uncorrelated case in (3.65).

If the CSI is not known at the transmitter and the MIMO channels are i.i.d. Rayleigh, then [91]

$$\lim_{K \to \infty} \frac{C^{\mathrm{BC}} \left(\{ \mathbf{H}_k^H \}, P/\sigma^2 \right)}{\log \log K} = 0, \tag{3.66}$$

again highlighting the importance of CSI at the transmitter for MU-MIMO.

As K increases asymptotically, the TDMA sum rate scales linearly with $\min(M, N)$ and double-logarithmically with respect to K due to multiuser diversity [86]

$$\lim_{K \to \infty} \frac{C^{\mathrm{TDMA}} \left(\left(\{ \mathbf{H}_k^H \}, P/\sigma^2 \right) \right)}{\log \log K} = \min(M, N). \tag{3.67}$$

Combining (3.65) and (3.67), we have that the gain of multiuser transmission over the best single-user transmission is:

$$\lim_{K \to \infty} \frac{C^{\mathrm{BC}} \left(\{ \mathbf{H}_k^H \}, P/\sigma^2 \right)}{C^{\mathrm{TDMA}} \left(\{ \mathbf{H}_k^H \}, P/\sigma^2 \right)} = \frac{M}{\min(M, N)}. \tag{3.68}$$

Therefore multiuser transmission maintains an advantage as long as the number of transmitter degrees of freedom M is larger than the number of degrees of freedom per user N.

Figure 3.17 shows the sum rate of the $(4, (K, N))$ BC and TDMA versus K, for $N = 1$ and 4. From (3.68), the ratio of the sum rates approaches 4 and 1, respectively for $N = 1$ and 4. For the range of K considered, the TDMA sum rate achieves a significant fraction of the sum-rate capacity except for the case of high SNR and $N = 1$.

For the MAC, we consider a modified case where each user transmits with power P and has a single antenna. From (3.33), the sum-rate capacity is

$$C^{\text{MAC}}\left(\{\mathbf{h}_k\}, KP/\sigma^2\right) = \log_2 \det \left(\mathbf{I}_M + \frac{P}{\sigma^2} \sum_{k=1}^{K} \mathbf{h}_k \mathbf{h}_k^H\right). \tag{3.69}$$

Since the channels are i.i.d. Rayleigh, we have that

$$\lim_{K \to \infty} \frac{1}{K} \sum_{k=1}^{K} \mathbf{h}_k \mathbf{h}_k^H = \mathbf{I}_M \tag{3.70}$$

almost surely, as a result of the law of large numbers. Therefore, for large K,

$$C^{\text{MAC}}\left(\{\mathbf{h}_k\}, KP/\sigma^2\right) \approx \log_2 \det \left(\mathbf{I}_M + \frac{KP}{\sigma^2} \mathbf{I}_M\right) \tag{3.71}$$

$$\approx M \log_2 \left(\frac{KP}{\sigma^2}\right). \tag{3.72}$$

It follows that

$$\lim_{K \to \infty} \frac{C^{\text{MAC}}\left(\{\mathbf{h}_k\}, KP/\sigma^2\right)}{\log K} = M. \tag{3.73}$$

For the BC with single-antenna users ($N = 1$) and fixed total transmit power, recall from (3.65) that the sum rate scales as $M \log \log K$ as the number of users K goes to infinity. On the MAC, however, each user brings its own power of P so that the total transmitted power increases linearly with K. This leads to the faster scaling of sum rate as $M \log K$ when $K \to \infty$.

3.7 Practical considerations

In this chapter we have described the capacity-achieving strategies for the MAC and BC. These techniques are based on successive interference cancellation and dirty paper coding, respectively. We have so far assumed these techniques operate ideally. However, in reality, there are many practical challenges in implementation which we describe now.

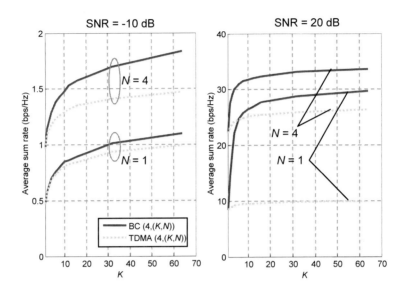

Fig. 3.17 Average sum rate versus K for $M = 4$. For asymptotically high K, the ratio of sum rates between BC and TDMA approaches $M/\min(M, N)$.

	MAC	BC	TDMA
	$\lim_{P/\sigma^2 \to 0} \dfrac{\bar{C}^{\mathrm{MAC}}((K,N), M, P/\sigma^2)}{P/\sigma^2}$ $= \rho(M, N) \log_2 \mathrm{e}$	$\lim_{P/\sigma^2 \to 0} \dfrac{\bar{C}^{\mathrm{BC}}(M, (K, N)P/\sigma^2)}{P/\sigma^2}$ $= \rho(M, N, K) \log_2 \mathrm{e}$	$\lim_{P/\sigma^2 \to 0} \dfrac{\bar{C}^{\mathrm{TDMA}}(M, (K, N), P/\sigma^2)}{P/\sigma^2}$ $= \rho(M, N, K) \log_2 \mathrm{e}$
	$\lim_{P/\sigma^2 \to \infty} \dfrac{\bar{C}^{\mathrm{MAC}}((K, N), M, P/\sigma^2)}{\log_2 P/\sigma^2}$ $= \min(M, KN)$	$\lim_{P/\sigma^2 \to \infty} \dfrac{\bar{C}^{\mathrm{BC}}(M, (K, N), P/\sigma^2)}{\log_2 P/\sigma^2}$ $= \min(M, KN)$	$\lim_{P/\sigma^2 \to \infty} \dfrac{\bar{C}^{\mathrm{TDMA}}(M, (K, N), P/\sigma^2)}{\log_2 P/\sigma^2}$ $= \min(M, N)$
	$\lim_{\substack{M, K \to \infty \\ K/M \to \alpha}} \dfrac{C^{\mathrm{MAC}}\left(\{\mathbf{h}_k\}, P/\sigma^2\right)}{M}$ $(N = 1) \quad = C^{\mathrm{Asym}}(\alpha, P/\sigma^2)$	$\lim_{\substack{M, K \to \infty \\ K/M \to \alpha}} \dfrac{C^{\mathrm{BC}}\left(\{\mathbf{h}_k\}, P/\sigma^2\right)}{M}$ $(N = 1) \quad = C^{\mathrm{Asym}}(\alpha, P/\sigma^2)$	$\lim_{\substack{M, K \to \infty \\ K/M \to \alpha}} \dfrac{C^{\mathrm{TDMA}}((\{\mathbf{h}_k^H\}, P/\sigma^2)}{M}$ $(N = 1) \quad = 0$
	$\lim_{K \to \infty} \dfrac{C^{\mathrm{MAC}}\left(\{\mathbf{h}_k\}, KP/\sigma^2\right)}{\log K}$ $(N = 1) \quad = M$	$\lim_{K \to \infty} \dfrac{C^{\mathrm{BC}}\left(\{\mathbf{H}_k^H\}, P/\sigma^2\right)}{\log \log KN}$ $= M$	$\lim_{K \to \infty} \dfrac{C^{\mathrm{TDMA}}\left(\{\mathbf{H}_k^H\}, P/\sigma^2\right)}{\log \log K}$ $= \min(M, N)$

Fig. 3.18 Summary of sum-rate capacity scaling.

3.7.1 Successive interference cancellation

The capacity-achieving strategy for the MIMO MAC is difficult to implement in practice from both the transmitter and receiver perspectives. At the transmitter, the optimal transmit covariances for the $N > 1$ case are difficult to compute because they require knowledge of all users' MIMO channels. One could do this by estimating channels at the central receiver, computing the covariances, and communicating them to the users via control channels. This strategy becomes challenging if the channels vary in time.

As a result of imperfect channel estimation, interference cancellation will not be ideal. Even if a stronger user is decoded perfectly, errors in the channel estimates for that user will lead to imperfect cancellation and therefore residual interference for weaker users. This becomes especially significant when the SNR difference between the two users is high, and the problem is worse with more users. In principle, partial interference cancellation with an MMSE criterion could be used to mitigate this effect [95].

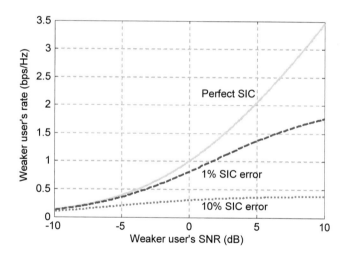

Fig. 3.19 Effect of cancellation error using SIC when detecting $K = 2$ users. The stronger user has an SNR of 15 dB higher. As a result of the SIC error, the performance of the weaker user is degraded.

For example, if we assume a 23 dB SNR difference between just two users, then even a 1% error in cancelling the stronger user's signal will result in a -3 dB cap on the SINR that the weaker user can achieve. In practice, this will mean that users with very disparate SNRs cannot be multiplexed, and this will also affect the spectral efficiency gain that successive interference cancellation can achieve. For illustration, Figure 3.19 shows the weaker user's rates for perfect cancellation, 1% cancellation error, and 10% cancellation error, assuming the weaker user's SNR is between -10 dB and 10 dB and the stronger user is always 15 dB higher. The main point to note is the ceiling imposed on the weaker user's rate, especially evident in the 10% case.

Another issue is the sample resolution in the receiver's analog-to-digital (A-to-D) converter in order to handle the dynamic range of the received signal. This can be modeled quite simply using results from rate-distortion theory. If the average power of the received signal (including noise) is P, and the A-to-D uses R bits/sym for quantization, then the quantization error can be modeled as an additional independent Gaussian noise of variance $P \times 2^{-R}$. Naturally, this added noise will affect the weaker user more than the stronger user.

3.7.2 Dirty paper coding

Dirty paper coding is difficult to implement due to the need for highly complex multidimensional vector quantization [96] [97] [98]. The time-varying nature of wireless channels makes the implementation even more difficult. Dirty paper coding with single-dimensional quantization can be implemented using Tomlinson-Harashima precoding, but there are significant performance losses for the low SINR regime of wireless communication. DPC also requires ideal channel state information at the transmitter which could be difficult to acquire in practice. This problem is discussed in Section 4.4 in the context of suboptimal broadcast channel techniques which also require CSIT.

3.7.3 Spatial richness

We recall that in SU-MIMO channels, multiple antennas are required at both the transmitter and receiver to achieve spatial multiplexing. However, if the antennas at either the transmitter or receiver array are fully correlated, then spatial multiplexing is not possible (Section 2.4). In a BC with M antennas and K single-antenna users, spatial multiplexing can be achieved even if the

M transmit antennas are fully correlated. For example if the channel is line-of-sight, then a user's channel vector would be determined by its direction relative to the array elements (2.79). If the K users are sufficiently separated in the azimuthal dimension, their channel vectors would be linearly independent, and it would be possible to achieve full multiplexing gain $\min(M, K)$. Furthermore, if each user had N fully correlated antennas, then multiplexing gain M could be achieved as long as the users are spatially separated and $K \geq M$. Even if the users' antennas were uncorrelated, the multiplexing gain would be the same: $M = \min(M, KN)$. Similar arguments can be made regarding the multiplexing gain of the MAC with fully correlated receive antennas.

3.8 Summary

In this chapter, we studied two canonical multiuser channels: the MIMO multiple access channel (MAC) and the MIMO broadcast channel (BC), both with additive white Gaussian noise (AWGN). The MIMO MAC models several users transmitting to a single base station (uplink), and the MIMO BC a single base station transmitting to several users (downlink).

- We examined the capacity regions and capacity-achieving techniques for these channels, assuming perfect CSI everywhere. On the MIMO MAC, receiver-side linear MMSE beamforming combined with sequential decoding and interference cancellation of user signals is capacity-achieving. On the MIMO BC, transmitter-side linear precoding combined with sequential encoding through dirty paper coding (DPC) is capacity-achieving.

- We stated an elegant and very useful duality result relating the MIMO MAC and MIMO BC, which allows us to translate BC problems to equivalent MAC problems. Briefly, the duality result states that the capacity region of the MIMO BC equals that of a dual MIMO MAC in which the channel matrices are all conjugate-transposed and the users are allowed to share power (with the sum power being the same as on the BC).

- We described algorithms for maximizing a weighted sum rate on the MIMO MAC as well as on the MIMO BC. On the MAC, the optimal rate vector always corresponds to decoding the users sequentially, in the order of increasing weight (so that the user with the highest weight does not see any interference). In contrast, on the BC, the optimal rate vector corresponds to encoding the users using dirty paper coding in the order of decreasing weight (so that the user with the lowest weight sees no interference). The algorithm for the BC exploits MAC-BC duality.

- Finally, we investigated the sum-rate performance achievable on both channels, and its asymptotic behavior with respect to SNR, number of antennas, and number of users. We also compared the sum-rate capacity to the sum rate achievable with simple TDMA. These results are summarized in Figure 3.18.

Chapter 4
Suboptimal Multiuser MIMO Techniques

The previous chapter described the capacity regions of multiuser MIMO channels and techniques for achieving capacity. Due to the complexity of these optimal techniques, we are motivated to consider suboptimal multiuser MIMO techniques with lower complexity.

For the multiple-access channel, we describe suboptimal beamforming strategies for mobiles with multiple antennas and alternatives to the capacity-achieving MMSE-SIC detector. For the broadcast channel, in order to avoid the complexity of implementing dirty paper coding, we describe linear precoding techniques which require only conventional single-user channel coding. These suboptimal precoding techniques can be classified according to the knowledge available at the transmitter: full channel state information, partial channel state information, statistical information, and no explicit information.

In the last section, we describe a duality for the MAC and BC under the assumptions of linear processing and single-antenna users. This duality is similar to the one for the general MIMO MAC and BC capacity regions and will be useful for the system performance evaluation in Chapter 6.

4.1 Suboptimal techniques for the multiple-access channel

In certain regimes that are relevant for cellular networks, the optimal capacity-achieving techniques for the MIMO MAC can be simplified. In particular, we recall that in the regime of high SNR for i.i.d. Rayleigh channels, optimal sum-capacity scaling proportional to $\min(M, KN)$ can be achieved without CSIT by having all K users transmitting isotropically over N antennas. If $K \geq M$, then optimal scaling of order M could be achieved by having each user transmit with just a single antenna. Therefore, while the capacity-achieving requirements for the MIMO broadcast channel and MIMO multiple-access channel are similar (both require ideal CSIT, linear precoding, ideal CSIR, and MMSE-SIC detection at the receiver), achieving near-optimal MIMO MAC capacity is generally regarded as simpler in practice because the requirement for CSIT is relaxed.

Transmitting a single stream per user is also optimal for the case of a large number of users, as we describe below. We also describe some lower complexity alternatives to using MMSE-SIC detection.

4.1.1 Beamforming for the case of many users

In general, the computation of the optimal precoder for each user in the MIMO MAC depends on the channels of all users, and a particular user could transmit multiple data streams. However, for a fixed M and N, as the number of users K grows without bound, transmitting a single stream per user with beamforming is both sufficient and necessary for achieving the sum-capacity [99]. In other words, if the number of users is large, multiple antennas at the mobile transmitter should be used for concentrating power in a single stream rather than for spatial multiplexing. Unfortunately, this result does not indicate how the beamforming should be performed.

For the related problem of achieving a fixed rate R/K for K users with minimum total transmit power, it was shown in [100] that a simple strategy of transmitting on each user's dominant eigenmode is optimal for fixed M and N as K increases asymptotically. This strategy is myopic in the sense that user k requires knowledge of only its channel $\mathbf{H}_k$. The covariance for user k is $P_k \mathbf{v}_k \mathbf{v}_k^H$, where P_k is the transmit power and $\mathbf{v}_k$ is the dominant eigenvector of $\mathbf{H}_k$. The sum-capacity can be achieved using a matched-filter receiver followed by successive interference cancellation, and the transmit power for each user is determined iteratively given the transmitter and receiver strategies.

Based on the observations in [99] and [100] and recalling that full-power transmission is optimal when using an MMSE-SIC receiver, we consider the sum-rate performance of transmitting full power on each user's dominant eigenmode and using an MMSE-SIC receiver. Recall that the capacity-achieving transmit covariance for each user is in general dependent on all users' channels. Therefore this alternative is simpler not only because each user transmits only a single stream but because each requires knowledge of only its own channel. Figure 4.1 shows the sum-rate performance averaged over i.i.d. channels, compared with the sum-capacity for $N = 4$ transmit antennas per user, an SNR of 10 dB per user, and $M = 4$ receive antennas. As K grows, the proposed strategy's sum rate approaches the sum-capacity. We conjecture that the strategy is asymptotically optimal, with the intuition being that as the number of users increases, interference for any user appears spatially white such that myopic transmission without channel prewhitening should be optimal.

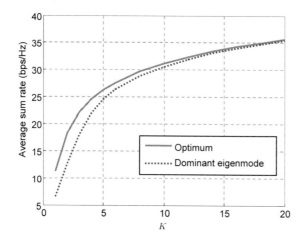

Fig. 4.1 Sum rate versus K for $M = 4$ and $N = 4$, average SNR per user is 10 dB. The suboptimal beamforming strategy achieves a significant fraction of the optimal sum rate, even for moderate K.

4.1.2 Alternatives for MMSE-SIC detection

The capacity-achieving MMSE-SIC detector is an example of *multiuser detection*, which refers to the class of techniques for detecting data signals in multiple-access channels. The fundamentals of multiuser detection are given in [43], and more recent advances in the field are described in [101]. Earlier results in multiuser detection were often developed in the context of code division multiple-access (CDMA) channels. In contrast to the MIMO multiple-access channel where users' signals are modulated by the channel $\mathbf{H}_k$, the users' signals in CDMA channels are modulated by different spreading sequences known to the receiver. The received signal model for the MIMO MAC (3.1) is applicable for CDMA channels, and multiuser detection techniques developed for these channels can therefore often be applied to the MIMO MAC.

If the channel state information at the MIMO MAC receiver is not reliable, successive interference cancellation may actually degrade performance, as discussed in Section 3.7.1. In this case, the linear MMSE receiver (without cancellation)(Section 2.3.1) is simpler alternative to the MMSE-SIC and achieves optimal multiplexing gain in the high-SNR regime for the SU-MIMO channel. Because of the equivalence between the open-loop (K, M) SU-MIMO channel and the $((K, 1), M)$ MIMO MAC, it also achieves optimal sum-rate scaling for the MAC. We can further extend this result to state that the MMSE receiver achieves optimal sum-rate scaling for the $((K, N), M)$ MIMO MAC if users transmit isotropically.

We note that if an MMSE receiver is used, full power transmission would not necessarily be optimal for weighted sum-rate maximization, as is the case for the MMSE-SIC. We can see this by considering a $((2, 1), M)$ MAC where both users have high SNR with $\mathbf{h}_1 = \mathbf{h}_2$. In this case, the user with the higher QoS weight should transmit and the other user should be silent. In general, optimizing the transmit powers for the MMSE receiver to maximize the sum rate is a non-convex optimization because of the interference. (In Section 4.3, an algorithm is given for determining the minimum sum of transmitted powers required to meet a set of SINR requirements for all users in a $((K, 1), M)$ MAC.)

If the noise power dominates over the interference power, then the MMSE-SIC provides minimal benefit over a simple matched filter. This situation applies to three of the quadrants in Figure 3.11 (high SNR with nearly or-

thogonal channels, and low SNR for both nearly orthogonal and highly correlated channels). In the fourth quadrant (high SNR with correlated channels),
TDMA transmission achieves a significant fraction of the MAC capacity region and also full multiplexing gain for the SU-MIMO channel. Even in cases
with moderate SNR, TDMA transmission may provide better performance
than simultaneous transmission and an MMSE receiver if the channels are
highly correlated.

4.2 Suboptimal techniques for the broadcast channel

Figure 4.2 shows an overview of suboptimal techniques for the broadcast
channel. The top branch distinguishes between single-user TDMA transmission and spatially multiplexed transmission for multiple users. We recall from
Section 3.6 that for asymptotically low SNR, TDMA transmission is actually
optimal. At high SNR, TDMA achieves multiplexing gain of $\min(M, N)$. Because the multiplexing gain of DPC is $\min(M, KN)$, TDMA achieves optimal
scaling if the number of antennas at each mobile is large ($N \geq M$). In general, time multiplexing of single-user transmission (using transmit diversity
or spatial multiplexing) could be used as a suboptimal strategy. However,
when the SNR is not too low and when the number of antennas per mobile is
small ($N < M$), we can do significantly better than TDMA using multiuser
transmission.

Under the multiuser transmission branch, we will focus on *linear precoding*
shown in Figure 4.3. If user $k = 1, ..., K$ transmits N_k independent data
streams, its information bits are channel encoded to create N_k-dimensional
data vectors $\mathbf{u}_k \in \mathbb{C}^{N_k \times 1}$. The encoding is done independently for each user,
so linear precoding is less complex than DPC. For a given symbol period, the
transmitted signal $\mathbf{s}$ is a linear combination of the user's coded data symbols:

$$\mathbf{s} = \sum_{k=1}^{K} \mathbf{G}_k \mathbf{u}_k, \qquad (4.1)$$

where $\mathbf{G}_k \in \mathbb{C}^{M \times N_k}$ is the precoding matrix for user k. In the case of a
rank-one covariance matrix ($N_k = 1$), linear precoding is often known as
beamforming. In this case, the transmitted signal is

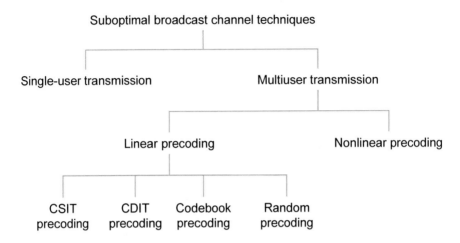

Fig. 4.2 Overview of suboptimal broadcast channel techniques.

$$\mathbf{s} = \sum_{k=1}^{K} \mathbf{g}_k u_k = \mathbf{Gu}, \tag{4.2}$$

where $\mathbf{G} := [\mathbf{g}_1 \ldots \mathbf{g}_K]$ and $\mathbf{u} = [u_1 \ldots u_K]^T$. Because a mobile can demodulate at most N spatially multiplexed streams, single-antenna mobiles can demodulate beamformed signals with $N_k = 1$, but not precoded signals with $N_k > 1$. For simplicity we will focus on the special case of beamforming but often use the more general "precoding" terminology.

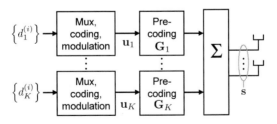

Fig. 4.3 A linear precoding architecture for suboptimal downlink MU-MIMO transmission. The data stream of user $k = 1, ..., K$ is channel encoded, and the data symbols $\mathbf{u}_k$ are precoded with the matrix $\mathbf{G}_k$.

Linear precoding techniques can be categorized into four classes depending on the knowledge available at the transmitter and the nature of the precoding matrices $\mathbf{G}_k$ (or beamforming vectors $\mathbf{g}_k$). These classes are described briefly here and treated in more detail in the remainder of the chapter.

- **CSIT precoding**: Knowledge of the channel state information (CSI) $\mathbf{H}_1^H, \ldots, \mathbf{H}_K^H$ is assumed at the transmitter. It is used when transmitter CSIT is highly reliable, for example in TDD systems. Obtaining CSIT is addressed in Section 3.4.

- **CDIT precoding**: Knowledge of the channel distribution information (CDI), from which channel realizations $\mathbf{H}_1^H, \ldots, \mathbf{H}_K^H$ are drawn, is assumed at the transmitter. CDIT precoding would be used when it is not possible to obtain reliable CSIT, for example in TDD systems with very fast fading channels or in FDD systems where information at the transmitter is obtained through an uplink feedback channel.

- **Codebook precoding**: Precoding matrices $\mathbf{G}_k$ are drawn from a codebook $\mathcal{B}$ which consists of a fixed set of matrices. This is a generalization of the single-user precoding technique described in Section 2.6. Like CDIT precoding, codebook precoding is used when it is not possible to obtain reliable CSIT. It is often used in FDD systems and is implemented by having a user transmit on an uplink feedback channel an index into the codebook to indicate its preferred precoding matrix.

- **Random precoding**: Precoding vectors are formed randomly, typically with no explicit knowledge of the channels at the transmitter and with minimal feedback from the mobiles. Random precoding is useful in analyzing the performance of networks with a large number of users. However, because quality of service requirements are difficult to meet, it is typically not used in practice.

We will see that linear precoding techniques like CSIT precoding can achieve a signficant fraction of the optimal sum-rate capacity if the number of users K is large relative to M and if an effective user- selection algorithm is used. For cases where K is small and where the CSI is known at the transmitter, *nonlinear* precoding techniques have been shown to provide performance gains over linear techniques by modifying the user data symbols nonlinearly to optimize the spatial characteristics of the transmitted signal. For example, under vector perturbation [102, 103] the transmitted signal can be written as $\mathbf{x} = \mathbf{G}(\mathbf{u} + \tau\mathbf{v})$, where $\mathbf{G}$ is a regularized zero-forcing matrix and $\mathbf{u}$ is the vector of coded symbols. The data vector is perturbed by a factor $\tau\mathbf{v}$, where τ is positive real number and $\mathbf{v}$ is a K-dimensional complex vector. These quantities are designed so that the perturbed data is nearly orthogonal to the eigenvectors associated with the large singular values of $\mathbf{G}$. Another nonlinear precoding technique is a generalization of Tomlinson-Harashima precoding [97]. A comparison of linear and nonlinear precoding techniques is given in [104].

In the following sections, we describe the four classes of linear precoding techniques in more detail. While each technique has different assumptions about the transmitter knowledge, each mobile receiver is assumed to have ideal knowledge of its own channel for demodulation.

4.2.1 CSIT precoding

Let us consider an $(M, (K, 1))$ BC using linear beamforming. The received signal by the kth user is

$$x_k = \mathbf{h}_k^H \mathbf{s} + n_k, \tag{4.3}$$

where the transmitted signal $\mathbf{s}$ is given by (4.2) and the corresponding SINR is

$$\Gamma_k = \frac{|\mathbf{h}_k^H \mathbf{g}_k|^2 v_k^2}{\sigma^2 + \sum_{j \neq k} |\mathbf{h}_k^H \mathbf{g}_j|^2 v_j^2}, \tag{4.4}$$

where $v_k = \mathbb{E}\left[|u_k|^2\right]$ $(k = 1, \ldots, K)$ is the power allocated to the kth user. The total transmit power is

$$tr\mathbb{E}[\mathbf{s}\mathbf{s}^H] = tr\left[(\mathbf{G}^H \mathbf{G})\mathbb{E}[\mathbf{u}\mathbf{u}^H]\right] \tag{4.5}$$

$$= \sum_k ||\mathbf{g}_k||^2 v_k. \tag{4.6}$$

We can define an optimization problem for determining the beamforming vectors $\mathbf{g}_k$ and user powers v_k $(k = 1, ..., K)$ to maximize the weighted sum rate subject to a transmit power constraint P. Given the quality of service weights q_k and channels $\mathbf{h}_k$, the optimization problem is

$$\max_{\mathbf{g}_k, v_k} \sum_{k=1}^K q_k \log_2 \left(1 + \frac{|\mathbf{h}_k^H \mathbf{g}_k|^2 v_k}{\sigma^2 + \sum_{j=1, j \neq k}^K |\mathbf{h}_k^H \mathbf{g}_j|^2 v_j} \right) \tag{4.7}$$

$$\text{subject to } v_k \geq 0, \quad k = 1, \ldots, K$$

$$\sum_k ||\mathbf{g}_k||^2 v_k \leq P.$$

In general, this is a non-convex optimization that can be solved using a branch-and-bound method [105]. Unfortunately, this technique is computationally very intensive and cannot be implemented in real-time. (The related problem of determining the minimum transmit power to meet a set of rate requirements is discussed below in Section 4.3.)

The optimization problem (4.7) can be simplified by imposing a *zero-forcing* constraint such that each user $k = 1, \ldots, K$ receives no interference from the signals of the other users:

$$|\mathbf{h}_k^H \mathbf{g}_j| = 0, j \neq k. \tag{4.8}$$

To meet this constraint, it is sufficient for $K \leq M$ and for the channels $\mathbf{h}_k$, $k = 1, \ldots, K$ to be linearly independent. In other words, defining $\mathbf{H} := [\mathbf{h}_1 \ldots \mathbf{h}_K]$ to be the $M \times K$ matrix of conjugate MISO channels, we require $K \leq M$ and the columns of $\mathbf{H}$ to be linearly independent. Then by defining the $M \times K$ precoding matrix to be

$$\mathbf{G} := \mathbf{H}(\mathbf{H}^H \mathbf{H})^{-1}, \tag{4.9}$$

the K-dimensional received signal vector $\mathbf{y} := [y_1 \ldots y_K]^T$ is

$$\mathbf{y} = \mathbf{H}^H \mathbf{x} + \mathbf{n} \tag{4.10}$$
$$= \mathbf{H}^H \mathbf{G} \mathbf{u} + \mathbf{n} \tag{4.11}$$
$$= \mathbf{u} + \mathbf{n}. \tag{4.12}$$

The zero-forcing criterion is satisfied because $\mathbf{H}^H \mathbf{G} = \mathbf{I}_K$. (One could balance the effects of interference and thermal noise using a *regularized* zero-forcing precoding, which is analogous to the MMSE receiver for the MAC [102].) Under zero-forcing, the kth user's signal is received with SNR v_k / σ^2, and its achievable rate is simply $\log(1 + v_k / \sigma^2)$. Using (4.9), the weighted sum-rate optimization (4.7) can be rewritten as:

$$\max_{v_k} \sum_{k=1}^{K} q_k \log\left(1 + \frac{v_k}{\sigma^2}\right) \tag{4.13}$$
$$\text{subject to } v_k \geq 0, \quad k = 1, \ldots, K$$
$$\sum_{k} \left[(\mathbf{H}^H \mathbf{H})^{-1}\right]_{(k,k)} v_k \leq P,$$

where the power constraint follows from (4.9).

In the case where $q_1 = \ldots = q_k$, the problem is equivalent to maximizing the sum rate over parallel Gaussian channels, and waterfilling [37] gives the optimal power allocation across the channels. A straightforward generalization of the waterfilling algorithm allows us to solve (4.13) when the QoS weights are different.

The optimization problem in (4.13) can be generalized in a number of ways that are useful for cellular networks. *User selection* addresses the problem of zero-forcing precoding when the number of users K exceeds the spatial degrees of freedom M. In practice, when antennas are powered by individual amplifiers, a *per-antenna power constraint* is more meaningful than the sum-power constraint we have assumed. We are also interested in zero-forcing-based transmission strategies for the case of *mobiles with multiple antennas*, enabling the possibility of multiplexing multiple streams for each user. These three extensions are addressed in the remainder of the section.

User selection

In deriving the zero-forcing precoding weights in (4.9), we ensured the existence of $(\mathbf{H}^H \mathbf{H})^{-1}$ by requiring $K \leq M$. In practice, the number of users K could exceed M for networks with a moderate density of users. In order to generalize (4.13) for the case where $K > M$, we let $\mathcal{U} \subseteq \{1, \ldots, K\}$ denote the subset of users during an interval that are candidates for service. Defining $|\mathcal{U}|$ to be the cardinality of $\mathcal{U}$, we restrict $|\mathcal{U}| \leq M$. For a given subset $\mathcal{U}$, we renumber the user indices and denote $\mathbf{H}(\mathcal{U}) := \left[\mathbf{h}_1 \ldots \mathbf{h}_{|\mathcal{U}|} \right]$ to be the $M \times |\mathcal{U}|$ matrix of MISO channels. We can rewrite the optimization (4.13) for a given $\mathcal{U}$ as follows:

$$\max_{v_k} \sum_{k=1}^{|\mathcal{U}|} q_k \log \left(1 + \frac{v_k}{\sigma^2} \right) \tag{4.14}$$

$$\text{subject to } v_k \geq 0, \quad k = 1, \ldots, |\mathcal{U}|$$

$$\sum_k \left[(\mathbf{H}(\mathcal{U})^H \mathbf{H}(\mathcal{U}))^{-1} \right]_{(k,k)} v_k \leq P.$$

Letting $R(\mathcal{U})$ be the weighted sum-rate solution to (4.14), the overall user-selection problem maximizes $R(\mathcal{U})$ over all possible subsets $\mathcal{U}$:

$$\max_{\mathcal{U} \subseteq \{1, \ldots, K\}} R(\mathcal{U}). \tag{4.15}$$

(We note that for maximizing the weighted sum rate using DPC, the two-level optimization is not necessary because the numerical algorithm in Section

3.5.2 automatically sets the covariances of inactive users to zero.) One could solve (4.15) by performing the optimization (4.14) over all possible subsets $\mathcal{U}$ and choosing the best subset. Using such a brute-force search, we would consider each of the K users individually, all pairs of users, all triplets of users, and so on up to all M-tuplets of users. The total number of possible subsets would be

$$\sum_{j=1}^{\min(M,K)} \frac{K!}{(K-j)!j!}. \qquad (4.16)$$

The number of possibilities could be quite large for even modest K. For example, with $M = 10$ and $K = 20$, we would need to consider over 100,000 subsets.

A greedy user-selection algorithm [106] can be used as a simpler alternative to the brute-force method. The algorithm successively adds users in a greedy manner to a candidate set of users. It starts by selecting the user which would achieve the highest weighted rate if the transmitter were to serve only this user. This user index is placed in the candidate set $\mathcal{U}$. The algorithm then calculates the weighted sum rate of all $K - 1$ pairs that include this user. If the maximum weighted sum rate among the pairs does not exceed the weighted rate of the single user, we exit the algorithm with $\mathcal{U}$ containing the single user. Otherwise, we let $\mathcal{U}$ be the winning pair and try adding another user. The algorithm proceeds iteratively in this manner until we run out of either spatial degrees of freedom or users.

Greedy user-selection algorithm

1. Initialization
   Set $n := 1$, $T_n := \emptyset$, DONE $:= 0$
2. While $n \leq \min(M, K)$ and DONE $== 0$, do the following:
   a. $k^* := \arg\max_{k \notin T_n} R\left(T_n \cup \{k\}\right)$
   b. if $R\left(T_n \cup \{k^*\}\right) < R\left(T_n\right)$
      $\mathcal{U} := T_n$
      DONE $:= 1$
   else
      $T_{n+1} := T_n \cup \{k^*\}$
      $n := n + 1$
   end

A related heuristic user-selection algorithm based on the user channel correlations rather than the user rates was proposed in [107]. Defining $\bar{R}^{\mathrm{ZF}}(M, (K, 1), P/\sigma^2)$ to be the average sum rate for zero-forcing using this selection algorithm, it was shown that this strategy achieves optimal sum-rate scaling (3.65) as K increases:

$$\lim_{K \to \infty} \frac{\bar{R}^{\mathrm{ZF}}(M, (K, 1), P/\sigma^2)}{\log \log K} = M \tag{4.17}$$

Per-antenna power constraints

So far we have assumed a sum-power constraint (SPC) that limits the total power summed across the M antennas. In practice, each transmit antenna is powered by a separate amplifier, and therefore a power constraint applied to each antenna would be more relevant. Under a *per-antenna power constraint* (PAPC), the transmit power of antenna m can be written as $\mathbb{E}\left[|\mathbf{s}_m|^2\right] = \sum_{k=1}^{K}|g_{m,k}|^2 v_k$, where $g_{m,k}$ is the (m, k)th element of the precoding matrix $\mathbf{G}$. Hence under a power constraint P_m for the mth antenna, the optimization (4.13) becomes:

$$\max_{v_k} \sum_{k=1}^{K} q_k \log\left(1 + \frac{v_k}{\sigma^2}\right) \tag{4.18}$$

$$\text{subject to } v_k \geq 0, \quad k = 1, \ldots, K$$

$$\sum_{k}|g_{m,k}|^2 v_k \leq P_m, \quad m = 1, \ldots, M.$$

Figure 4.4 shows the power feasibility region for a $K = 2$-user, $M = 2$-antenna case. The two axes correspond to the power v_k given to each of the users. For the PAPC, the feasible region is bounded by the linear power constraints imposed by the two antennas and by the positive axes. Each antenna is assumed to have a power limit of $P/2$. The feasibility region for the SPC P is bounded by the positive axes and a single linear power constraint. The PAPC region is contained within the sum power region, and the boundaries of the two regions intersect at the vertex of the PAPC region. The optimal operating point is the point lying on the feasibility region boundary which is tangent to the sum-rate contour curve with the highest value. This problem is a convex optimization which can be solved using conventional numerical optimization techniques [108].

Figure 4.5 shows the sum-rate performance for a system with $M = 4$ transmit antennas serving $K = 4$ or 20 users, each with $N = 1$ antenna.

We consider three variations of ZF: with a sum-power constraint (SPC) and greedy selection, with a per-antenna power constraint (PAPC) and brute-force selection, and with PAPC and greedy selection. We compare zero-forcing performance with optimal sum-rate capacity achieved using DPC and with the TDMA transmission strategy where the user with the highest closed-loop capacity is served (Section 3.6).

At very low SNR, TDMA is optimal. Therefore for lower SNR, the gap between the sum-rate capacity and TDMA sum rate is smaller. At asymptotically high SNR, TDMA achieves a multiplexing gain of $\min(M, N) = 1$ while the optimal multiplexing gain of the broadcast channel is $\min(M, KN) = 4$. The spatial multiplexing advantage of the zero-forcing techniques is clear at higher SNRs because the slope of all three ZF curves (for a given value of K) is the same as the sum-capacity slope, while the TDMA slope is much shallower.

In going from $K = 4$ to 20 users, the ZF performance improves because it is more likely to find more favorable channel realizations as the number of users increases. However, the slope of the curves at high SNR is not dependent on K because the multiplexing gain is always 4. For $K = 4$ users under a per-antenna power constraint, brute-force ZF provides a slight performance advantage over greedy ZF. However, for $K = 20$, there is visually no difference between the two, and we show only a single curve for the PAPC performance.

Zero-forcing for $N > 1$: BD and MET

If the receivers have multiple antennas $N > 1$, then the extra degrees of freedom could be used to demodulate multiple data streams. Let us first consider the case where each user demodulates $N_k = N$ streams and N is such that $M = NK$. The received signal by user k is

$$\mathbf{x}_k = \mathbf{H}_k^H \mathbf{s} + \mathbf{n}_k = \mathbf{H}_k^H \sum_{j=1}^{K} \mathbf{G}_j \mathbf{u}_j + \mathbf{n}_k, \qquad (4.19)$$

where $\mathbf{G}_k \in \mathbb{C}^{M \times N}$ and $\mathbf{u}_k \in \mathbb{C}^N$ are, respectively, the precoding matrix and data vector for user k. A generalization of the ZF technique for this case is known as *block diagonalization* (BD) [85, 109], where the precoding matrices are designed so that $[\mathbf{H}_1 \ldots \mathbf{H}_K]^H [\mathbf{G}_1 \ldots \mathbf{G}_K] \in \mathbb{C}^{KN \times KN}$ has a block-diagonal structure consisting of K blocks of size $N \times N$ along the diagonal and zeros elsewhere. Each user receives N streams with no interference from the other users.

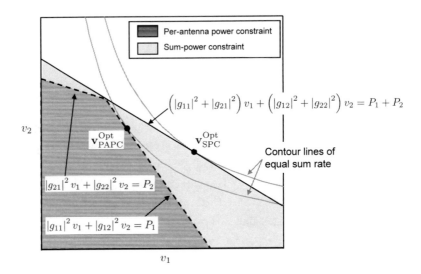

Fig. 4.4 The shaded regions show the allowable power allocations under a sum-power constraint and a per-antenna power constraint for $K = 2$ users and $M = 2$ transmit antennas. The operating points for maximizing the sum rate under PAPC and SPC are highlighted.

Under BD, the precoding matrix for the kth user is the product of two matrices: $\mathbf{G}_k = \mathbf{G}_k^{(L)} \mathbf{G}_k^{(R)}$. The left matrix $\mathbf{G}_k^{(L)}$ is size $M \times N$ and lies in the null space of the other users' MIMO channels: $\mathbf{H}_j^H \mathbf{G}_k^{(L)} = \mathbf{0}_{(N \times N)}, j \neq k$. The right matrix $\mathbf{G}_k^{(R)}$ is size $N \times N$ and is the right unitary matrix of the singular value decomposition (SVD) of $\mathbf{H}_k^H \mathbf{G}_k^{(L)}$. In other words, given the decomposition

$$\mathbf{H}_k^H \mathbf{G}_k^{(L)} = \mathbf{V}_k^{(L)} \mathbf{\Lambda}_k \mathbf{V}_k^{(R)H}, \tag{4.20}$$

we assign $\mathbf{G}_k^{(R)} := \mathbf{V}_k^{(R)}$. User k uses $\mathbf{V}_k^{(L)H} \in \mathbb{C}^{N \times N}$ as the linear combiner for its received signal. As a result of the block-diagonal criterion, the combiner output given the received signal (4.19) is

$$\begin{aligned}
\mathbf{V}_k^{(L)H} \mathbf{x}_k &= \mathbf{V}_k^{(L)H} \mathbf{H}_k \mathbf{G}_k^{(L)} \mathbf{G}_k^{(R)} \mathbf{u}_k + \mathbf{V}_k^{(L)H} \mathbf{n}_k \\
&= \mathbf{V}_k^{(L)H} \mathbf{V}_k^{(L)} \mathbf{\Lambda}_k \mathbf{V}_k^{(R)H} \mathbf{V}_k^{(R)} \mathbf{u}_k + \mathbf{V}_k^{(L)H} \mathbf{n}_k \\
&= \mathbf{\Lambda}_k \mathbf{u}_k + \mathbf{V}_k^{(L)H} \mathbf{n}_k.
\end{aligned}$$

Because $\mathbf{V}_k^{(L)}$ is a unitary matrix, the resulting noise vector is spatially white since $\mathbf{n}_k$ is spatially white. Waterfilling can be used to allocate power across

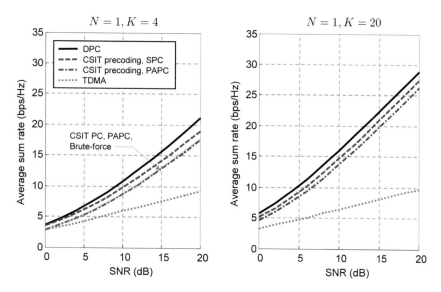

Fig. 4.5 Average sum rate versus SNR for $M = 4$, $K = 4, 20$, $N = 1$. The average sum-rate capacity for the broadcast channel (achieved using DPC) is compared with CSIT pre-coding (zero-forcing) and TDMA transmission. Three variations of zero-forcing are shown: sum-power constraint (SPC), per-antenna power constraint (PAPC) with brute-force user selection, and PAPC with greedy-user selection. As SNR increases, all three zero-forcing sum-rate curves have the same slope as the sum-rate capacity, indicating optimal scaling with respect to M. With only a single-antenna per user, single-user transmission with TDMA cannot achieve multiplexing gain.

the N data streams of $\mathbf{u}_k$. In doing so, the kth user achieves the closed-loop MIMO capacity for the equivalent single-user MIMO channel $\mathbf{H}_k^H \mathbf{G}_k^{(L)}$. The BD technique is a generalization of both ZF precoding to the case of multiple receive antennas and closed-loop single-user MIMO to the case of multiple users.

The basic BD technique that we have just described is designed for the specific case of $M = KN$ where we transmit exactly N streams to each user. Even though the transmitter could send up to M streams, this strategy may not maximize the resulting sum rate. For example, if the channels between users are highly correlated, imposing a ZF constraint could be too restrictive, and the sum-rate performance could be improved by sending fewer streams.

A generalization of BD known as *multiuser eigenmode transmission* (MET) [110] maximizes the weighted sum rate by distributing up to M streams among the K users so that each user receives *up to N* streams. (In general, each user can receive up to $\min(N, M)$ streams, but we will assume that $N \leq M$.) Unlike BD, the number of streams sent to each user is not necessarily N. MET is based on the SVD of each user's channel $\mathbf{H}_k^H$, characterized by its N singular values. The base transmits up to N streams to each user in the direction of its eigenmodes, and it uses a zero-forcing criterion so that its users experience no interference.

We first describe the precoding and receiver combining for maximizing the MET weighted sum rate for a given set of active users and active eigenmodes. We let $\mathcal{U}$ be the set of active users. Given the SVD of $\mathbf{H}_k^H$, $k \in \mathcal{U}$, we let $\mathcal{E}_k \subseteq \{1, \ldots, N\}$ be the set of indices for its active eigenmodes. If $|\mathcal{E}_k|$ is the number of active eigenmodes for the kth user, the total number of eigenmodes must satisfy $\sum_{k \in |\mathcal{U}|} |\mathcal{E}_k| \leq M$. We let $\mathbf{\Lambda}_k$ be the $|\mathcal{E}_k| \times |\mathcal{E}_k|$ diagonal matrix whose components are the singular values of $\mathbf{H}_k^H$ corresponding to $\mathcal{E}_k$. We let $\mathbf{V}_k^{(L)}$ and $\mathbf{V}_k^{(R)}$ be the corresponding submatrices of the left and right unitary SVD matrices. To simplify notation, we do not explicitly write out the dependence of $\mathbf{\Lambda}_k$, $\mathbf{V}_k^{(L)}$, and $\mathbf{V}_k^{(R)}$ on $\mathcal{E}_k$. We have that

$$\underbrace{\mathbf{V}_k^{(L)H}(\mathcal{E}_k)}_{|\mathcal{E}_k| \times N} \underbrace{\mathbf{H}_k^H}_{N \times M} = \underbrace{\mathbf{\Lambda}_k(\mathcal{E}_k)}_{|\mathcal{E}_k| \times |\mathcal{E}_k|} \underbrace{\mathbf{V}_k^{(R)H}(\mathcal{E}_k)}_{|\mathcal{E}_k| \times M}. \tag{4.21}$$

As in the BD formulation, the precoding matrix for user k can be decomposed into a left and right component: $\mathbf{G}_k = \mathbf{G}_k^{(L)} \mathbf{G}_k^{(R)}$. The matrix $\mathbf{G}_k^{(L)}$ for MET spans the null space of $\mathbf{V}_j^{(L)H} \mathbf{H}_j^H$ for $j \in \mathcal{U}, j \neq k$, and it is of size

$$M \times (M - \sum_{j \in |\mathcal{U}|, j \neq k} |\mathcal{E}_j|).$$

Unlike the BD design where $\mathbf{G}_k^{(L)}$ needed to be orthogonal to all N dimensions of the other users' MIMO channels, $\mathbf{G}_k^{(L)}$ for MET is orthogonal to the $|\mathcal{E}_j| \leq N$ selected eigenmode dimensions of the other active users j.

Processing the received signal $\mathbf{x}_k$ with the $|\mathcal{E}_k| \times N$ combiner $\mathbf{V}_k^{(L)H}$ and relying on the zero-forcing properties of $\mathbf{G}_k^{(L)}$, the combiner output is

$$\mathbf{V}_k^{(L)H} \mathbf{y}_k = \mathbf{V}_k^{(L)H} \mathbf{H}_k^H \mathbf{G}_k^{(L)} \mathbf{G}_k^{(R)} \mathbf{u}_k + \tag{4.22}$$
$$\sum_{j \in \mathcal{U}, j \neq k} \mathbf{V}_k^{(L)H} \mathbf{H}_k^H \mathbf{G}_j^{(L)} \mathbf{G}_j^{(R)} \mathbf{u}_j + \mathbf{V}_k^{(L)H} \mathbf{n}_k$$
$$= \mathbf{\Lambda}_k \mathbf{V}_k^{(R)H} \mathbf{G}_k^{(L)} \mathbf{G}_k^{(R)} \mathbf{u}_k + \mathbf{V}_k^{(L)H} \mathbf{n}_k.$$

We decompose the $|\mathcal{E}_k| \times (M - \sum_{j \in |\mathcal{U}|, j \neq k} |\mathcal{E}_j|)$ matrix $\boldsymbol{\Lambda}_k \mathbf{V}_k^{(R)H} \mathbf{G}_k^{(L)}$, and we write the SVD as

$$\boldsymbol{\Lambda}_k \mathbf{V}_k^{(R)H} \mathbf{G}_k^{(L)} = \left[\tilde{\mathbf{V}}_k^{(L)} \right] \left[\tilde{\boldsymbol{\Lambda}}_k \quad \mathbf{0} \right] \left[\tilde{\mathbf{V}}_k^{(R)H} \right], \qquad (4.23)$$

where $\tilde{\boldsymbol{\Lambda}}_k$ is the diagonal matrix of $|\mathcal{E}_k|$ singular values. Letting the matrix $\mathbf{G}_k^{(R)H}$ be the top $|\mathcal{E}_k| \times (M - \sum_{j \in |\mathcal{U}|, j \neq k} |\mathcal{E}_j|)$ submatrix of $\tilde{\mathbf{V}}_k^{(R)H}$, post-multiplying the combiner output in (4.22) with $\tilde{\mathbf{V}}_k^{(L)}$ results in a $|\mathcal{E}_k|$-dimensional vector:

$$\begin{aligned}
\tilde{\mathbf{V}}_k^{(L)H} \mathbf{V}_k^{(L)H} \mathbf{y}_k &= \tilde{\mathbf{V}}_k^{(L)H} \boldsymbol{\Lambda}_k(\mathcal{E}_k) \mathbf{V}_k^{(R)H} \mathbf{G}_k^{(L)} \mathbf{G}_k^{(R)} \mathbf{u}_k + \mathbf{n}_k' \\
&= \tilde{\mathbf{V}}_k^{(L)H} \left[\tilde{\mathbf{V}}_k^{(L)} \right] \left[\tilde{\boldsymbol{\Lambda}}_k \quad \mathbf{0} \right] \left[\tilde{\mathbf{V}}_k^{(R)H} \right] \mathbf{G}_k^{(R)} \mathbf{u}_k + \mathbf{n}_k' \\
&= \tilde{\boldsymbol{\Lambda}}_k \mathbf{u}_k + \mathbf{n}_k',
\end{aligned}$$

where processed noise vector $\mathbf{n}_k' := \tilde{\mathbf{V}}_k^{(L)H} \mathbf{V}_k^{(L)H} \mathbf{n}_k$ is spatially white.

The eigenmodes for all active users are decoupled, and the power can be allocated using waterfilling to maximize the weighted sum rate. If we denote $v_{k,j} := \mathbb{E}\left[|u_{k,j}|^2 \right]$ to be the power allocated to the jth eigenmode ($j \in \mathcal{E}_k$) of user k ($k \in \mathcal{U}$), the transmitted signal for user k is $\mathbf{G}_k \mathbf{u}_k$, and the power transmitted for user k is $tr\mathbf{G}_k^H \, \mathrm{diag}[v_{k,1} \ldots v_{k,|\mathcal{E}_k|}]\mathbf{G}_k$. For a given set of active users $k \in \mathcal{U}$ and their active eigenmodes $\mathcal{E}_k$, the weighted sum-rate optimization is

$$\max_{v_{k,l}} \sum_{k \in \mathcal{U}} q_k \sum_{l \in \mathcal{E}_k} \log_2 \left(1 + \frac{\tilde{\lambda}_{k,l}^2 v_{k,l}}{\sigma^2} \right) \qquad (4.24)$$

$$\text{subject to } v_{k,l} \geq 0, \quad k \in \mathcal{U}, l \in \mathcal{E}_k$$

$$\sum_{k \in \mathcal{S}} tr\mathbf{G}_k^H \, \mathrm{diag}[v_{k,1} \ldots v_{k,|\mathcal{E}_k|}]\mathbf{G}_k \leq P.$$

Similar to the optimization for the original ZF technique (4.14), the optimization in (4.24) can be solved using weighted waterfilling over the $\sum_{k \in \mathcal{U}} |\mathcal{E}_k|$ parallel channels. Letting $R(\mathcal{U}, \{\mathcal{E}_k\})$ be the solution to (4.24) for a given active set $\mathcal{U}$ and active eigenmodes $\mathcal{E}_k$ for $k \in \mathcal{U}$, the outer optimization that maximizes the weighted sum rate over all combinations of eigenmodes and active user sets is

$$\max_{\mathcal{U} \subseteq \{1, \ldots, K\}, \mathcal{E}_k, k \in \mathcal{U},} R(\mathcal{U}, \{\mathcal{E}_k\}). \qquad (4.25)$$

To solve this optimization problem, one could resort to a brute-force search as was done for the ZF case.

Alternatively, a generalization of the the user selection algorithm described for ZF could be used for MET, where the selection occurs over KN virtual users corresponding to the KN eigenmodes. Under MET, at most M streams are transmitted and distributed among the K users so that an active user receives at most $\min(M, N)$ streams. All the eigenmodes could be assigned to a single user, or they could be distributed among multiple users. Therefore closed-loop SU-MIMO is a special case of MET, and the optimization in (4.25) automatically chooses between single-user and multiuser MIMO transmission on a frame-by-frame basis.

Single-stream MET

If multiple users are served under MET each with a single stream, it is in general not necessary for each user to be served on its dominant eigenmode (i.e., the mode with the largest singular value). The reason is that the dominant eigenmodes of the users could be unfavorably correlated. However, the likelihood of these unfavorable correlations decreases as the number of users K increases. It is shown in [111] that transmitting on the dominant eigenmode for M out of K users is indeed optimal in the sense that the resulting sum rate and the sum-rate capacity approach zero as K increases. Therefore, in a similar way that single-stream transmission is asymptotically optimal for the MAC (Section 4.1) for large K, single-stream transmission (beamforming) is also optimal for the BC.

The MET optimization in (4.25) can be modified for single-stream transmission by restricting the candidate eigenmode set $\mathcal{E}_k$ for the kth user to contain at most a single eigenmode. We call this technique *MET-1*. The single eigenmode is not necessarily restricted to be the dominant one because for smaller K it is useful to allow this flexibility. Compared to MET, MET-1 reduces complexity in three ways. It reduces the control information required to notify an active user which eigenmode to demodulate, it reduces the user/eigenmode selection complexity, and it reduces the receiver complexity since multi-stream demodulation is not necessary.

Figure 4.6 shows the transmission options for various transceiver strategies for a $(4,(2,2))$ broadcast channel. The eigenmodes of user 1 are labeled 1A and 1B, and those for user 2 are 2A and 2B. Eigenmodes 1A and 2A are the users' respective dominant eigenmodes. Under single-user transmission, the transmitter chooses among the following four options for maximizing the weighted sum-rate metric: {1A},{2A},{1A,1B},{2A,2B}. Under BD, the base activates all four eigenmodes. MET has the most options to consider.

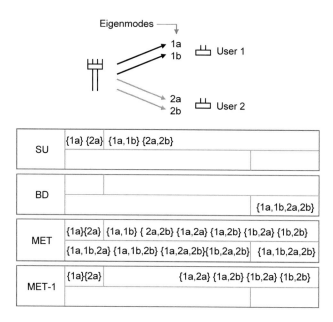

Fig. 4.6 Transmission options for a $(4, (2, 1))$ BC using linear precoding. User 1 has eigenmodes 1A and 1B, and user 2 has eigenmodes 2A and 2B. Potential active eigenmode sets are shown for different transceiver options. SU-MIMO serves either user 1 or 2. BD transmits on all eigenmodes. MET is the most general and considers all possible eigenmode combinations. MET-1 transmits at most one stream to each active user.

Because of the spatial interaction among eigenmodes, transmission on non-dominant eigenmodes could be optimal. For example, if 1A and 2A are highly correlated, the maximum sum rate could be achieved by activating 1B and 2B. Under MET-1, the transmission options are reduced. If the number of users is small, as in this example, the sum-rate performance of MET-1 could suffer significantly. However, as K increases, the average sum-rate performance of MET-1 approaches that of MET, as shown in Figure 4.7.

Figure 4.8 shows the sum-rate performance versus SNR for $M = 4$, $K = 4, 20$, and $N = 4$. The sum-rate capacity (achieved with DPC), MET, MET-1, and best single-user MIMO transmission are considered. Compared to Figure 4.5, where the mobiles had only a single antenna, all techniques have the same multiplexing order 4, resulting in the same slope for high SNR. For $K = 4$, there is a significant performance gap between MET and MET-1,

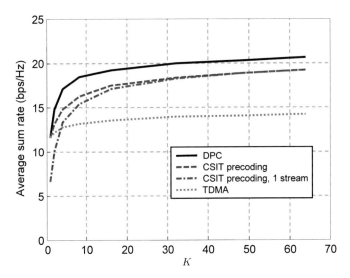

Fig. 4.7 Sum-rate performance versus the number of users K, for $M = 4$, $N = 4$, $SNR =$ 10 dB. CSIT precoding using MET achieves a significant fraction of the sum-capacity. As the number of users increases, CSIT precoding restricted to one stream (MET-1) also achieves a significant fraction. Even though SU-MIMO achieves the full multiplexing order of 4, the MU-MIMO options achieve higher sum rate because of the greater flexibility in distributing streams among users.

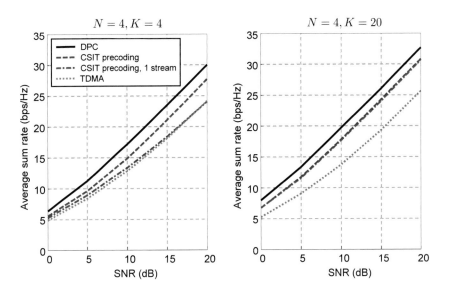

Fig. 4.8 Sum rate versus SNR for $M = 4$, $K = 4, 20$, $N = 4$. MET achieves a significant fraction of sum-capacity. Even though all three techniques have the same spatial degrees of freedom, MET performs better than SU-MIMO due to multiuser diversity when $K > M$.

indicating that single-stream transmission is far from ideal. For high SNR, TDMA performance approaches that of MET-1. If there are more users ($K = 20$), the gap between MET and MET-1 is negligible over the range of SNRs, and there is a significant advantage over TDMA.

4.2.2 CDIT precoding

CDIT precoding for the broadcast channel assumes that at the transmitter, the distributions of the K users' channels are known but the realizations of the channels are not known. The problem is to determine the precoders and power allocations for the K users to maximize the *expected* weighted sum rate. From (4.7), the required optimization is

$$\max_{\mathbf{g}_k, v_k} \sum_{k=1}^{K} q_k \mathbb{E}\left[\log_2\left(1 + \frac{|\mathbf{h}_k^H \mathbf{g}_k|^2 v_k}{\sigma^2 + \sum_{j=1, j \neq k}^{K} |\mathbf{h}_k^H \mathbf{g}_j|^2 v_j}\right)\right] \quad (4.26)$$
$$\text{subject to } v_k \geq 0, \quad k = 1, \ldots, K$$
$$\sum_k \|\mathbf{g}_k\|^2 v_k \leq P.$$

For the special case of a single-user ($K = 1$) MISO channel, the optimal precoding vector $\mathbf{g}_1$ corresponds to the eigenvector of $\mathbb{E}[\mathbf{h}_1 \mathbf{h}_1^H]$, and full power P is applied in this direction [112]. For the single-user MIMO channel, the optimal transmit covariance $\mathbf{Q}_1$ for maximizing the average capacity can be decomposed as $\mathbf{Q}_1 = \mathbf{V} \mathbf{P} \mathbf{V}^H$, where the eigenvectors of $\mathbf{Q}$ are the columns of the unitary matrix $\mathbf{V}$ and where the eigenvalues are given by the diagonal entries of $\mathbf{P} = \text{diag}\{p_1, \ldots, p_M\}$. The eigenvectors of the capacity-achieving transmit covariance equal those of $\mathbb{E}[\mathbf{H}_1 \mathbf{H}_1^H]$, and the eigenvalues can be found through a numerical algorithm to satisfy a set of necessary and sufficient conditions [112]. Because the parallel channels are not orthogonal, the power allocation does not correspond to a waterfilling solution.

For the MISO BC with multiple $K > 1$ users, the general solution for maximizing the expected weighted sum rate (4.26) is not known. Some ad hoc techniques have been proposed, including a variation of zero-forcing based on channel statistics [113] and a technique combining CDIT and knowledge of the instantaneous channel norm [114].

For the special case of a line-of-sight channel, the users' directions can be determined from the CDIT. Therefore knowledge of the channel $\mathbf{h}_k$ can be obtained implicitly from 2.79 (modulo the phase offset) and CSIT precoding could be implemented. If ZF precoding is deployed, up to $\min(M, K)$ users could be served on non-interfering beams.

4.2.3 Codebook precoding

In Section 2.3.7, we discussed codebook (or limited feedback) precoding for the single-user MIMO channel. The receiver estimates the MIMO channel and feeds back B bits to the transmitter as an index to indicate its preferred precoding matrix chosen from a codebook containing 2^B precoding matrices (codewords). For the multiuser case, the codebook is known by the K users, and each user feeds back B bits to indicate its preferred codeword. A block diagram for codebook precoding for multiuser MISO channels is shown in Figure 4.9.

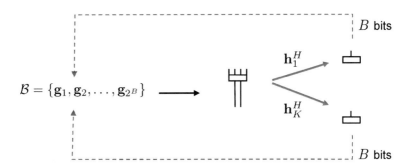

Fig. 4.9 Block diagram for multiuser codebook precoding. User k estimates its channel $\mathbf{h}_k^H$ and feeds back B bits to indicate its perfered precoding vector. The codebook $\mathcal{B}$ consists of 2^B codeword matrices and is known by the transmitter and the user.

In designing codebooks for single-user transmission, we recall that codewords are designed to closely match the realizations or eigenmodes of the channel. This principle can be applied in the multiuser case. However, when serving multiple users, the interaction of the precoding matrices between users needs to be considered. While the transmitter could serve a user with its requested codeword, it could also decide to use another codeword because the requested one would create too much interference for the other users.

We consider a $(M, (K, 1))$ broadcast channel and suppose for now that we restrict the precoding vector to one drawn from the codebook $\mathcal{B}$ consisting of M-dimensional codewords each with unit norm. To maximize the sum rate, the precoding vectors for the K users $\mathbf{g}_1, \ldots, \mathbf{g}_K \in \mathcal{B}$ and power allocations $P_1, \ldots, P_K$ need to be determined:

$$\max_{\sum_k P_k \leq P} \max_{\mathbf{g}_1, \ldots, \mathbf{g}_K \in \mathcal{B}} \sum_{k=1}^{K} \log_2 \left(1 + \frac{|\mathbf{h}_k^H \mathbf{g}_k|^2 P_k}{\sigma^2 + \sum_{j \neq k, j=1}^{K} |\mathbf{h}_k^H \mathbf{g}_j|^2 P_j} \right). \qquad (4.27)$$

This is a non-convex optimization which is difficult to solve. By further restricting the power allocation among users to be equal, the optimal assignment of codewords could be obtained through an exhaustive search if $\left| \mathbf{h}_k^H \mathbf{g}_j \right|$ is known for all k and j.

Alternatively, if equal power allocation is assumed and if the kth user knew in advance the precoding vectors of the other $K-1$ users $\mathbf{g}_j, j = 1, \ldots, K, j \neq k$, it would be straightforward to determine its optimal precoding vector:

$$\mathbf{g}_k^{\text{Opt}} = \arg \max_{\mathbf{g} \in \mathcal{B}} \log_2 \left(1 + \frac{|\mathbf{h}_k^H \mathbf{g}|^2 P/K}{\sigma^2 + \sum_{j \neq k, j=1}^{K} |\mathbf{h}_k^H \mathbf{g}_j|^2 P/K} \right). \qquad (4.28)$$

If there are K precoding vectors in the codebook, a user could determine its best precoding vector using (4.28) if it assumes that for any vector it picks, the other $K-1$ vectors will be active.

PU2RC

This last strategy is the basis of *per-user unitary rate control* (PU2RC), in which the precoding vectors are the columns of $2^B/M$ (assumed to be an integer) unitary matrices of size $M \times M$. During a given transmission interval, the active codewords are restricted to come from the columns of the same unitary matrix. Therefore for a given candidate vector $\mathbf{g}$ drawn from the unitary matrix u ($u = 1, \ldots, 2^B/M$), the mobile knows that the other $M-1$ active vectors are the other columns of this unitary matrix. It can therefore use (4.28) to choose the best among 2^B codewords.

If the number of users K is large compared to the number of codewords 2^B, this strategy works well because it will likely find M users to serve who are nearly orthogonal. However, if the number of users is small, their desired vectors may be sparsely distributed among the columns of the unitary matrices, and the number of users selecting any particular matrix would be less than M. As a consequence, for a fixed K, the PU2RC sum rate decreases as the size of the codebook 2^B increases [115]. This result is somewhat counterintuitive since one would expect performance to improve as the knowledge of the channel becomes more refined with increasing B.

Line-of-sight channels

In line-of-sight channels for single-user transmission, an effective codebook could consist of 2^B codewords that generate directional beams based on MRT. Under multiuser transmission, the beamforming vectors could be redesigned to reduce the interbeam interference compared to the MRT beams. Returning to the example in Section 2.3.7.1, suppose we want to serve a $120°$ sector by creating four beams in directions $45°, 75°, 105°, 135°$ using a linear array with M elements in Figure 2.17. If we partition the azimuthal dimension into four $30°$ sectors, the ideal beam response would be unity within the desired range and zero outside this range. For example, the beam pointing in the direction $\theta^* = 105°$ would ideally have a unity response in the range $(105° \pm 15°)$ and zero elsewhere, as shown in Figure 4.10.

The beamforming design for a linear array in line-of-sight channels is essentially equivalent to the design of finite-impulse response (FIR) digital filters, where the angular domain in the former is analogous to the frequency domain of the latter [116]. Therefore filter design techniques for minimizing the frequency response in the stopband region can be used for minimizing the sidelobe levels. Figure 4.10 shows the directional response of a more sophisticated beamformer based on a Dolph-Chebyshev design criterion. It has significantly lower sidelobe levels with only a slightly wider main-lobe response compared to the MRT response used for single-user beamforming.

Codebook-based zero-forcing

In general, the transmitter is not restricted to use the codewords in the codebook $\mathcal{B}$. While the feedback from the users index these codewords, the actual precoding matrix for a given user could be a function of the selected codewords of all K users.

For example, for the case of single-antenna users, suppose each user selects the beamforming vector according to the single-user criterion (2.77). If the base treats the desired beamforming vector of each user as the Hermitian conjugate of its actual MISO channel, it could perform *codebook-based zero-forcing precoding* as described in Section 4.2.1. For $K \leq M$, we

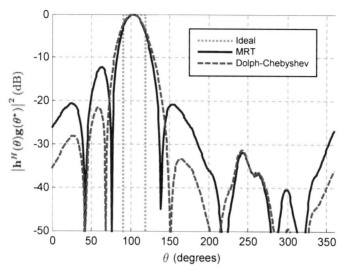

Fig. 4.10 The directional response for a beamforming vector **g** pointing in the direction $\theta^* = 105°$. The ideal beam pattern response has unity gain over the directional range of interest and zero gain outside this range. The beamformer **g** based on maximal ratio transmission is matched to the channel $\mathbf{h}(\theta^*)$. A more sophisticated beamformer based on a Dolph-Chebyshev design has lower sidelobes, resulting in less interbeam interference for users lying outside the range of interest.

would define $\mathbf{H} := [\mathbf{g}_1 \dots \mathbf{g}_K]$ and use the precoding matrix defined in (4.9):
$\mathbf{G} := \mathbf{H} \left(\mathbf{H}^H \mathbf{H} \right)^{-1}$.

A performance analysis of codebook-based zero-forcing in i.i.d. channels is given in [117] using random codebooks, where 2^B codewords are randomly placed on the surface of an M-dimensional hypersphere. It was shown that as the average SNR of the users increases, the number of feedback bits B must be increased linearly with the SNR (in decibels) in order to achieve the full multiplexing gain in an $(M, M, 1)$ broadcast channel. This is in contrast to the single-user MIMO link where no feedback is necessary at any SNR to achieve full multiplexing gain. By modifying the metric for determining the codebook feedback, one can improve the sum-rate performance for a given number of feedback bits B. For example in [118], the feedback is chosen so the resulting SINR is robust against all possible beam assignments at the base. Generalizations of codebook-based zero-forcing precoding to the case of multiple antenna receivers $N > 1$ are described in [60, 119, 120].

4.2.4 Random precoding

Under random precoding, beams are formed randomly instead of being drawn from a predetermined codebook. These techniques work well when the number of users K vying for service is large. We first consider the case where a single beam is formed to serve a single user at a time. We then consider the more general case where M random mutually orthogonal beams are formed to serve M users.

Single-beam

The concept was first proposed for the purpose of achieving multiuser diversity by artificially inducing SNR fluctuations through random beamforming [121]. During frame t, a base equipped with M antennas creates a random unit-power M-dimensional vector $\mathbf{g}(t)$ drawn from an isotropic distribution. Each user measures and feeds back its SNR

$$\frac{P}{\sigma^2}|\mathbf{h}_k^H \mathbf{g}(t)|^2, \tag{4.29}$$

where the channel $\mathbf{h}_k$ is an i.i.d. complex Gaussian channel and does not vary with t. The base uses proportional scheduling, serving the user that achieves the maximum weighted rate where the weight is the reciprocal average rate:

$$k^*(t) = \arg \max_k \frac{1}{\bar{R}_k(t)} \log_2 \left(1 + \frac{P}{\sigma^2}|\mathbf{h}_k^H \mathbf{g}(t)|^2 \right). \tag{4.30}$$

Under proportional fair scheduling, the long-term average rate as t approaches infinity can be shown to exist, and each user is served a fraction $1/K$ of the frames. If the number of users K is very large, the average rate achieved by user k is:

$$\lim_{K \to \infty} \left[K \lim_{t \to \infty} \bar{R}_k(t) \right] = \log_2 \left(1 + \frac{P}{\sigma^2}||\mathbf{h}_k||^2 \right). \tag{4.31}$$

In other words, its average rate is the same as if the base pointed a beam $\mathbf{g}(t) = \mathbf{h}_{k^*(t)}/||\mathbf{h}_{k^*(t)}||$ directly at the user everytime it was scheduled. This beam would have required full CSIT, but random opportunistic precoding achieves the same performance using only SNR feedback. The key is that random precoding is coupled with opportunistic scheduling according to (4.30). This performance gain would clearly not be achieved with random scheduling. With a large number of users, the randomly formed beam will be well-matched to some user during every frame. Even if the MISO channels had different statistics, this technique would still work as long as the random beams had matching statistics.

For correlated channels, $\mathbf{g}(t)$ creates a directional beam with a random direction, and instead of using a random sequence of beams, one could use an orderly cyclic scanning of the desired sector with fixed angular steps [121]. The beam sequence could be modified based on waiting times of users, resulting in improved throughput and packet delays for the case of finite user populations [122]. A shortcoming of using only a single beam is that this strategy does not achieve linear capacity scaling with M. To address this problem, we consider a generalization of the technique for multiple beams.

Multiple beams

As described in [93], a transmitter with M antennas creates a set of M random orthonormal beams $\mathbf{g}_1, \ldots, \mathbf{g}_M$ drawn from an isotropic distribution. User k measures its SINR over each of the M beams (assuming power P/M per beam) and feeds back its best SINR along with the corresponding beam index $m = 1, \ldots, M$:

$$b^* = \arg \max_{b \in \{1 \ldots M\}} \frac{\frac{P}{M\sigma^2} |\mathbf{h}_k^H \mathbf{g}_b|^2}{1 + \sum_{j=1, j \neq b}^{M} \frac{P}{M\sigma^2} |\mathbf{h}_k^H \mathbf{g}_j|^2}. \tag{4.32}$$

(The feedback overhead can be reduced by requiring feedback from only users whose best SINR exceeds a threshold which is independent of K [93].) The base transmits on M beams, where each beam is directed to the user with the highest SINR. The sum rate of opportunistic beamforming averaged over the channel realizations is given by:

$$\bar{R}^{\mathrm{OB}}(M, (K, 1), P/\sigma^2) =$$
$$\mathbb{E}_{\mathbf{h}_1 \ldots \mathbf{h}_K} \sum_{b=1}^{M} \log_2 \left(1 + \max_{i \in \{1 \ldots K\}} \frac{\frac{P}{M\sigma^2} |\mathbf{h}_i^H \mathbf{g}_b|^2}{1 + \sum_{j=1, j \neq b}^{M} \frac{P}{M\sigma^2} |\mathbf{h}_i^H \mathbf{g}_j|^2} \right). \tag{4.33}$$

As the number of users increases asymptotically, multi-beam opportunistic precoding achieves the optimal sum-rate scaling:

$$\lim_{K \to \infty} \frac{\bar{R}^{\mathrm{OB}}(M, (K, 1), P/\sigma^2)}{\log \log K} = M. \tag{4.34}$$

Intuitively, this strategy achieves linear scaling with M because it becomes increasingly likely as K increases to find a set of M users whose channels are orthogonal and match the M random beams. In fact, opportunistic precoding not only achieves optimal sum-rate scaling but actually achieves optimal sum rate asymptotically:

$$\lim_{K \to \infty} \left[\bar{R}^{\mathrm{OB}}(M, (K, 1), P/\sigma^2) - \bar{C}^{\mathrm{BC}}(M, (K, 1), P/\sigma^2) \right] = 0, \qquad (4.35)$$

where the average sum rate for the broadcast channel is defined in (3.45). Even though the sum rate scales optimally, there is a significant performance gap as seen in Figure 4.11. Strategies for narrowing the performance gap for sparse networks are described in [123].

Note that whereas the single-beam case considered scheduling over static channels for each user, the multibeam sum rate in (4.33) considers the maximum "one-shot" sum rate averaged over random realizations for each user. Under this formulation, the average sum rate does not rely on the orthonormal beams to be random. Therefore codebook precoding with an arbitrary fixed set of M orthonormal beams in the codebook achieves the same performance and requires the same feedback (SINR and beam index for each user) as random precoding.

In practice, we would be interested in the scheduled performance for a finite number of users. We can compare codebook precoding using an arbitrary set of M orthonormal beams that do not change in time with random precoding using a set of beams that change during each frame. If for a given set of K users the channels $\mathbf{h}_1, \ldots, \mathbf{h}_K$ each change randomly from frame to frame (with infinite doppler), then the underlying statistics of the SINR in (4.32) do not depend on whether the beams $\mathbf{g}_1, \ldots, \mathbf{g}_M$ are fixed in time or not. The scheduled performance is therefore equivalent for the two strategies.

On the other hand, if the user channels are fixed over time (zero doppler), then codebook precoding has a disadvantage because a user's channel could have a poor orientation with respect to the codebook vectors, and its misfortune does not change with time. With random precoding, however, the relationship of a user's channel with the random vectors changes each frame, allowing it to take advantage of favorable beam realizations. If hierarchical feedback is allowed for codebook precoding, the precoding vectors could be refined over time as described in Section 2.6, and its performance would surpass that of random precoding.

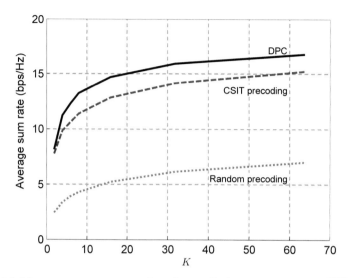

Fig. 4.11 Mean sum rate versus number of users K. Average user SNR is SNR = 20 dB. Random precoding requires no explicit CSI information at the transmitter and achieves the sum-rate capacity asymptotically as K increases. However its performance is far from optimal for reasonable K. On the other hand, ZF requires full CSIT but achieves a significant fraction of the sum-rate capacity for all K.

4.3 MAC-BC duality for linear transceivers

In this section, we will describe a duality between the $((K, 1), M)$ MAC and the $(M, (K, 1))$ BC that holds under the restriction of linear processing, i.e., when the MAC receiver and BC transmitter are constrained to implement only linear beamforming at their M antennas, in order to separate the signals among the K users. This means that successive interference cancellation is not permitted on the MAC, and dirty paper coding is not permitted on the BC. This duality is similar in spirit to the one described for the MAC and BC capacity regions (Section 3.4), and it will be used for the simulation methodology in Section 5.4.4.

For the purpose of stating the duality results, it will be convenient to normalize the noise power at each antenna to 1 in the MAC and BC channel models of (3.1) and (3.2). We will accordingly model the MAC by

$$\mathbf{x} = \sum_{k=1}^{K} \mathbf{h}_k s_k + \mathbf{n}, \tag{4.36}$$

where $\mathbf{n} \sim CN\left(\mathbf{0}, \mathbf{I}_M\right)$, and the BC by

$$x_k = \mathbf{h}_k^H \mathbf{s} + n_k, \quad k = 1, 2, \ldots, K, \tag{4.37}$$

where $n_k \sim CN(0,1)$ for each k.

On the MAC, suppose user k transmits at power p_k and is received with the unit-norm receiver beamforming vector $\mathbf{w}_k$. The resulting SINR for user k is then

$$\alpha_k = \frac{\left|\mathbf{w}_k^H \mathbf{h}_k\right|^2 p_k}{1 + \sum_{j \neq k} \left|\mathbf{w}_k^H \mathbf{h}_j\right|^2 p_j}. \tag{4.38}$$

Denote by Γ_{MAC} the set of all such achievable SINR vectors $(\alpha_1, \alpha_2, \ldots, \alpha_K)$, over all choices of the beamforming vectors $\mathbf{w}_1, \mathbf{w}_2, \ldots, \mathbf{w}_K$ and powers $p_1, p_2, \ldots, p_K$.

Turning our attention now to the BC, suppose the base station uses power q_k and the unit-norm transmitter beamforming vector $\mathbf{w}_k$ for user k. Then the resulting SINR for user k is

$$\beta_k = \frac{\left|\mathbf{h}_k^H \mathbf{w}_k\right|^2 q_k}{1 + \sum_{j \neq k} \left|\mathbf{h}_k^H \mathbf{w}_j\right|^2 q_j}. \tag{4.39}$$

Denote by Γ_{BC} the set of all such achievable SINR vectors $(\beta_1, \beta_2, \ldots, \beta_K)$, over all choices of the beamforming vectors $\mathbf{w}_1, \mathbf{w}_2, \ldots, \mathbf{w}_K$ and powers $q_1, q_2, \ldots, q_K$.

The duality result for linear processing, originally due to [124] and [125], states that the achievable SINR regions Γ_{MAC} and Γ_{BC} on the MAC and BC, respectively, are identical:

$$\Gamma_{MAC} = \Gamma_{BC}. \tag{4.40}$$

Further, any feasible SINR vector $(\gamma_1, \gamma_2, \ldots, \gamma_K)$ can be achieved with the *same* beamforming vectors and the *same* sum power in both channels, i.e., there must exist unit-norm beamforming vectors $\mathbf{w}_1, \mathbf{w}_2, \ldots, \mathbf{w}_K$, MAC powers $p_1, p_2, \ldots, p_K$, and BC powers $q_1, q_2, \ldots, q_K$ for the K users, such that

$$\frac{\left|\mathbf{w}_k^H \mathbf{h}_k\right|^2 p_k}{1 + \sum_{j \neq k} \left|\mathbf{w}_k^H \mathbf{h}_j\right|^2 p_j} = \gamma_k = \frac{\left|\mathbf{h}_k^H \mathbf{w}_k\right|^2 q_k}{1 + \sum_{j \neq k} \left|\mathbf{h}_k^H \mathbf{w}_j\right|^2 q_j} \tag{4.41}$$

and

$$\sum_{k=1}^{K} p_k = \sum_{k=1}^{K} q_k. \tag{4.42}$$

4.3.1 MAC power control problem

On the MAC, given feasible SINR targets $\gamma_1, \gamma_2, \ldots, \gamma_K$ for the K users, we wish to find transmitted powers $p_1, p_2, \ldots, p_K$ and unit-norm receiver beamforming vectors $\mathbf{w}_1, \mathbf{w}_2, \ldots, \mathbf{w}_K$ that minimize the sum of the transmitted powers while meeting all the SINR targets. More formally, we have the optimization problem

$$\min_{\{p_k\},\{\mathbf{w}_k\}} \sum_{k=1}^{K} p_k \tag{4.43}$$

subject to

$$\frac{\left|\mathbf{w}_k^H \mathbf{h}_k\right|^2 p_k}{1 + \sum_{j \neq k} \left|\mathbf{w}_k^H \mathbf{h}_j\right|^2 p_j} \geq \gamma_k, \quad \|\mathbf{w}_k\| = 1, \quad p_k \geq 0 \ \forall k. \tag{4.44}$$

The following simple iteration has been shown to solve the above problem [126], regardless of how the user powers are initialized:

MAC power control algorithm

1. Given powers $p_1^{(n)}, p_2^{(n)}, \ldots, p_K^{(n)}$ for the K users, let $\mathbf{w}_k^{(n)}$ be the unit-norm vector that maximizes user k's SINR:

$$\mathbf{w}_k^{(n)} = \frac{\mathbf{Q}_k^{(n)} \mathbf{h}_k}{\left\|\mathbf{Q}_k^{(n)} \mathbf{h}_k\right\|}, \quad \text{where } \mathbf{Q}_k^{(n)} = \left(\mathbf{I}_M + \sum_{j \neq k} p_j^{(n)} \mathbf{h}_j \mathbf{h}_j^H\right)^{-1}. \tag{4.45}$$

2. Update user k's power to just attain its target SINR of γ_k, assuming that every other user j continues to transmit at power $p_j^{(n)}$ and user k's receiver beamforming vector is $\mathbf{w}_k^{(n)}$:

$$p_k^{(n+1)} = \frac{\gamma_k \left(1 + \sum_{j \neq k} \left|(\mathbf{w}_k^{(n)})^H \mathbf{h}_j\right|^2 p_j^{(n)}\right)}{\left|(\mathbf{w}_k^{(n)})^H \mathbf{h}_k\right|^2}. \tag{4.46}$$

4.3.2 BC power control problem

On the BC, given feasible SINR targets $\gamma_1, \gamma_2, \ldots, \gamma_K$ for the K users, we wish to find transmitted powers $q_1, q_2, \ldots, q_K$ and unit-norm transmitter beamforming vectors $\mathbf{w}_1, \mathbf{w}_2, \ldots, \mathbf{w}_K$ that minimize the sum of the transmitted powers while meeting all the SINR targets. More formally, we have the optimization problem

$$\min_{\{q_k\}, \{\mathbf{w}_k\}} \sum_{k=1}^{K} q_k \tag{4.47}$$

subject to

$$\frac{\left|\mathbf{h}_k^H \mathbf{w}_k\right|^2 q_k}{1 + \sum_{j \neq k} \left|\mathbf{h}_k^H \mathbf{w}_j\right|^2 q_j} \geq \gamma_k, \quad \|\mathbf{w}_k\| = 1, \quad q_k \geq 0 \; \forall k. \tag{4.48}$$

The following simple iteration has been shown to solve the above problem [124], regardless of how the user powers are initialized:

BC power control algorithm

1. Given downlink powers $q_1^{(n)}, q_2^{(n)}, \ldots, q_K^{(n)}$ and virtual uplink powers $p_1^{(n)}, p_2^{(n)}, \ldots, p_K^{(n)}$ for the K users, let $\mathbf{w}_k^{(n)}$ be the unit-norm vector that maximizes user k's SINR on the virtual uplink:

$$\mathbf{w}_k^{(n)} = \frac{\mathbf{Q}_k^{(n)} \mathbf{h}_k}{\left\|\mathbf{Q}_k^{(n)} \mathbf{h}_k\right\|}, \quad \text{where } \mathbf{Q}_k^{(n)} = \left(\mathbf{I}_M + \sum_{j \neq k} p_j^{(n)} \mathbf{h}_j \mathbf{h}_j^H\right)^{-1}. \tag{4.49}$$

2. Update user k's downlink and virtual uplink powers to just attain its target SINR of γ_k on both the downlink and the virtual uplink, assuming that the powers for every other user j remain at $q_j^{(n)}$ (downlink) and $p_j^{(n)}$ (virtual uplink), and user k's beamforming vector is $\mathbf{w}_k^{(n)}$ on *both* the downlink and the virtual uplink:

$$p_k^{(n+1)} = \frac{\gamma_k \left(1 + \sum_{j \neq k} \left|(\mathbf{w}_k^{(n)})^H \mathbf{h}_j\right|^2 p_j^{(n)}\right)}{\left|(\mathbf{w}_k^{(n)})^H \mathbf{h}_k\right|^2}, \tag{4.50}$$

$$q_k^{(n+1)} = \frac{\gamma_k \left(1 + \sum_{j \neq k} \left|\mathbf{h}_k^H \mathbf{w}_j^{(n)}\right|^2 q_j^{(n)}\right)}{\left|\mathbf{h}_k^H \mathbf{w}_k^{(n)}\right|^2}. \tag{4.51}$$

4.4 Practical considerations

In this section, we describe challenges for implementing downlink precoding in practice, including the acquisition of CSIT and the auxiliary signalling such as pilot signals.

4.4.1 Obtaining CSIT in FDD and TDD systems

CSIT precoding requires ideal knowledge of the users' channel state information at the transmitter. In frequency-division duplexed (FDD) systems, the downlink and uplink transmissions occur on different bandwidth resources. Mobiles could estimate the CSI based on downlink pilot signals, quantize the estimate, and feed back this information to the base over an uplink channel. This is a special case of feedback for codebook precoding where the codewords designed to approximate the channel realizations. The quality of the CSI used by the transmitter would depend on the quality of the initial downlink estimate, the bandwidth of the uplink feedback channel, and the time variation of the channel.

Alternatively, one could use unquantized *analog feedback* by modulating an uplink control signal with the estimated channel coefficient. An analysis of zero-forcing beamforming with realistic channel estimation and both quantized and unquantized feedback is given in [127]. Its conclusion is that very significant downlink throughput is achievable with simple and efficient channel state feedback, provided that the feedback link over the uplink MAC is properly designed. A comprehensive survey of techniques for communicating under limited feedback is found in [128].

In time-division duplexed (TDD) systems, transmissions for the uplink and downlink are multiplexed in time over the same bandwidth. By estimating the user channels on the uplink (based on user pilot signals), the base could use these estimates for the following downlink transmission. If the users were transmitting data on the uplink, their channels would need to be estimated for coherent detection. Therefore by exploiting the channel reciprocity, the base could obtain CSI for downlink transmission with no additional overhead. However, because the uplink CSI estimates are based on uplink pilot channels, the pilot overhead increases as the number of users K and number of antennas per user N increases. For a fixed overhead, the reliability of the CSIT depends on the duplexing interval relative to the channel coherence time. Overall, ideal CSIT could be a reasonable assumption for a TDD system with a small number of slowly moving users.

In practice, calibration of the RF chains is also required to ensure channel reciprocity [129]. This procedure can be performed within a fraction of a second and relatively infrequently as the electronics drift (on the order of minutes), so it does not contribute significantly to the overhead.

While TDD-enabled CSIT results in better performance than an FDD system with limited feedback, the system implementation of TDD is often more restrictive because the downlink and uplink transmissions among all bases in the network must be time synchronized in order to avoid potentially catastrophic interference. In other words, all bases must transmit at the same time and receive at the same time. As seen in Figure 4.13 if the two bases are not time-synchronized, a downlink transmission by base A could cause severe interference for the uplink reception of base B. The interference power from base A could be significantly higher than the desired user's signal because the base station transmit power is much higher than the mobile power and because the channel between the two bases is potentially unobstructed. The received interference power could be high enough to overload the sensitive low-noise amplifiers at the front end of base B and to prevent any baseband detection or interference mitigation techniques from being used.

This synchronization requirement applies to base stations owned by different operators that share the same tower. Even though the different bases operate on different frequencies, the out-of-band interference could still be detrimental if the base stations are not synchronized. Therefore while TDD systems offer the potential of adjusting the duplexing fraction between uplink and downlink transmissions, this adjustment must be made on a network-wide basis in order to prevent severe interference.

4.4.2 Reference signals for channel estimation

Pilot signals (or *reference signals* as they are known in the 3GPP standards) are sent from the transmitter and used to enable CSI estimation at the receiver for coherent demodulation. There are two classes of reference signal structures, as shown in Figure 4.14.

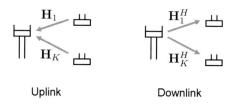

Uplink   Downlink

Fig. 4.12 For TDD systems, channel reciprocity is used to obtain CSIT. On the uplink, pilots are transmitted by the users, and the base estimates the channels $\mathbf{H}_1, \ldots, \mathbf{H}_K \in \mathbb{C}^{N \times M}$. On the downlink, the base uses the complex conjugate of each estimate for transmission.

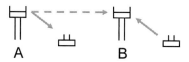

A   B

Fig. 4.13 Under TDD, if base stations A and B are not time-synchronized, a downlink transmission by base A will cause severe interference for the uplink reception of base B.

Using *common reference signals*, a reference signal $\mathbf{p}_m \in \mathbb{C}^T$ spanning T time symbols is transmitted on antenna $m = 1, \ldots, M$. The duration T is designed so that the channel is relatively static over these symbols. (In general, the reference signals can span both the time and frequency domains, for example, in OFDMA systems.) The reference signals are mutually orthogonal in the time dimension: $\mathbf{p}_m^H \mathbf{p}_n = 0$ for any $m \neq n$. Therefore the length of the reference signals must be at least as large as the number of antennas: $T \geq M$. Mobile k correlates the received signal with each of the known reference signals to obtain an estimate of the channel $\mathbf{h}_k$. The reference signals are "common" in the sense that they are utilized by all users.

Using *dedicated reference signals*, a reference signal is assigned to each beam. Figure 4.14 shows an example where dedicated reference signals transmitted for 2^B codewords in codebook $\mathcal{B}$. By correlating the received signal by each of the reference signals $\mathbf{p}_1, \ldots, \mathbf{p}_{2^B}$, the mobile obtains estimates of the beamformed channels $\mathbf{h}_k^H \mathbf{g}_1, \ldots, \mathbf{h}_k^H \mathbf{g}_{2^B}$. Dedicated reference signals are sometimes known as *user-specific reference signals* when a beam is assigned to a specific user. In this case, the assigned user correlates the received signal with only the associated reference signal assigned to it.

Dedicated reference signals benefit from the transmitter combining gain of the beamforming vector, resulting in a greater range for a fixed reference signal power compared to the common reference signals. For coherent demodulation of a signal modulated with precoding vector $\mathbf{g}$, user k requires an estimate of the equivalent channel $\mathbf{h}_k^H \mathbf{g}$. Dedicated reference signals *must* be used when the precoding vector is not known at the mobile, for example with CSIT or random precoding. Otherwise, if the precoding vector $\mathbf{g}$ is known, it could first estimate the channel $\mathbf{h}_k^H$ using the common reference signals and then compute the estimate for $\mathbf{h}_k^H \mathbf{g}$.

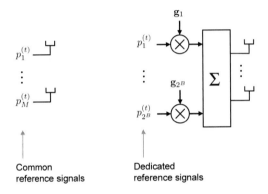

Fig. 4.14 Two types of reference (pilot) symbols. Common reference symbols allow mobile k to estimate its channel $\mathbf{h}_k^H$. Dedicated reference symbols allow mobile k to estimate the beamformed channels $\mathbf{h}_k^H \mathbf{g}_1, \ldots, \mathbf{h}_k^H \mathbf{g}_{2^B}$.

4.5 Summary

This chapter describes techniques for the MIMO MAC and BC channels that provide suboptimal performance but with lower complexity compared to capacity-achieving techniques.

- For the MAC, the design of optimal transmit covariances for users with multiple antennas is difficult. Beamforming (transmitting with a rank-one covariance) becomes optimal as the number of users increases asymptotically.

- For the BC, we focus on precoding techniques where linear transformations are applied to independently encoded user data streams. CSIT precoding is best suited to TDD systems where CSI is readily available at the base station transmitter through channel reciprocity. Zero-forcing beamforming is a type of CSIT precoding in which active users receive their desired data signal with no interference. CDIT and codebook precoding are suited to FDD systems where accurate CSIT is not available. Random precoding is suitable for systems with a large number of users.

- For the single-antenna MAC and BC constrained to linear processing, we also describe simple iterative algorithms for minimizing the sum power transmitted while attaining given feasible SINR targets for all the users. The algorithm for the MAC actually minimizes every user's power individually.

Chapter 5
Cellular Networks: System Model and Methodology

In the previous chapters we discussed MIMO techniques for isolated single-user and multiuser channels where the received signal is corrupted only by thermal noise. In this and the following chapter, we study MIMO techniques in cellular networks where the performance is corrupted by co-channel interference in addition to thermal noise. In this chapter, we describe a cellular network model and methodology for evaluating its performance by numerical simulation. The following chapter presents performance results and insights for MIMO system design.

Detailed performance evaluation of MIMO techniques in cellular networks is challenging due to the complexity of these networks. One would have to model each component of the system in great detail, accounting for practical attributes and impairments such as user mobility, spatial correlation, finite data buffers, channel estimation errors, hybrid ARQ, standards-dependent frame structure, data decoding, and errors in the control channels. Modeling these details in a simulation would result in accurate performance predictions, at a cost of great complexity and loss of generality.

At the other extreme, purely analytic techniques have been applied to a simplified cellular network model known as the Wyner model [130] where bases and users lie on a line instead of a plane. This one-dimensional model provides many fundamental insights of theoretical interest. However, their applicability to more practical cellular networks is limited because the model does not account for basic attributes such as shadowing and random user placement.

Hybrid semi-analytic techniques have also been proposed to bridge the gap between analysis and detailed simulations [131–134]. These techniques have wider applicability than the purely analytic ones, but they have limited flexibility because they apply only to specific antenna configurations.

The performance evaluation technique used in this book is based on Monte Carlo simulations, which capture key aspects of a generic cellular network. In order to focus on fundamental insights, we use performance metrics based on Shannon capacity described in earlier chapters and make simplifying assumptions which allow us to understand the *relative* performance of the various MIMO techniques. The methodology can be applied to different antenna architectures and accounts for system elements that are common to contemporary cellular networks including IEEE 802.16e, 802.16m, LTE, and LTE Advanced.

Section 5.1 provides a broader overview of the cellular network design problem described in Chapter 1. Section 5.2 describes the general system model for performance evaluation, including the model for sectorization, the reference SNR parameter, and a technique for accounting for interference at the edge of the network. In Section 5.3, two modes of base station operation are discussed along with various forms of spatial interference mitigation. Section 5.4 describes the performance metrics and the simulation methodology for the numerical simulation results in Chapter 6. Section 5.5 describes implementation aspects of contemporary packet-scheduled cellular systems, including sectorization, scheduling, and acquiring channel state information at the transmitter (CSIT).

Depending on the context, a *cell site* (or simply *site*) is defined to be either the hardware assets associated with a hexagonal cell or the hexagonal cell itself. Each cell site is partitioned into S sectors, and, unless otherwise noted, a base station with M antennas is associated with each sector.

5.1 Cellular network design components

A framework for cellular network design can be given as follows: for a given geographic area defined by the characteristics of its channels and expected data traffic, design the medium access control (MAC) and physical (PHY) layers, a base station architecture, a placement of base stations, and a terminal architecture to optimize the spectral efficiency per unit area, subject to resource and cost constraints. (From the context, it will be clear whether "MAC" stands for "medium access control" or "multiple-access channel".)

This framework is illustrated in Figure 5.1, and we consider each of the components below.

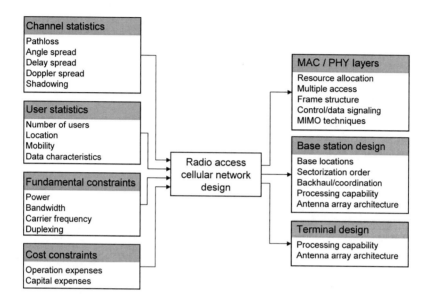

Fig. 5.1 Overview of the characteristics, constraints, and outputs for a radio access cellular network design.

Channel characteristics

The system should be designed to match the statistics of the channel, which depend on factors such as the terrain (urban, suburban, rural), the location of the users (indoors or outdoors), and the user characteristics. The channel is characterized by its distance-based pathloss, its shadowing characteristics, and its variations in time, frequency and space [39]. For the distance-based pathloss, we use a simple slope-intercept model [135] to describe the signal attenuation as a function of the distance traveled relative to the pathloss at a fixed distance given by the intercept value which is dependent on the carrier frequency. Whereas in free space the power decays with the square of distance, signals in most cellular channels decay with a coefficient closer to the model of an ideal reflecting ground plane whose value is 4. Channel modeling will be discussed in Section 5.2.

Variations in the time, frequency, and spatial domains are measured respectively by the doppler spread, delay spread, and angle spread. These parameters are dependent on the carrier frequency. In addition, the doppler spread is dependent on the mobile speed, the delay spread is dependent on the distance of scatterers relative to the base or mobile, and the angle spread is dependent on the relative height of the scatterers with the antennas. Therefore these channel statistics are dependent on the characteristics of the users and on the base station antenna array architecture. The shadowing of a channel captures longer-term characteristics of the channel that depend on the location of large-scale terrain features like buildings and mountains as well as on the large-scale density and location of scatterers.

User characteristics

The network provides service to a population of users that is characterized by a spatial distribution, a velocity distribution and service requirements. A realization of the spatial distribution gives the location for a user, and a realization of the velocity distribution gives its speed. Only a subset of the users are actively transmitting or receiving data while the rest of them are in an idle state. The service requirements for each user depend on the application, such as voice, data downloading, streaming media, or gaming. These characteristics can be parameterized by their average data rate and latency requirements.

Fundamental constraints

Fundamental constraints for cellular network design are based on regulations for using the licensed spectrum. An operator purchases bandwidth on certain carrier frequencies and is allowed to transmit with a limited effective radiated power. The spectrum is typically sold in paired bands for frequency-division duplexing (FDD) or unpaired bands for time-division duplexing (TDD).

Cost constraints

Monetary costs associated with network design include capital expenses for developing and deploying network infrastructure (base stations and backhaul) and operational expenses that include power, rent on real estate, maintenance, and backhaul [136]. The computational complexity of signal processing and network management algorithms has an impact on the cost of hardware and software implementation.

MAC and PHY layer design

The physical layer defines the specifications for the air interface, including the frame structure, options for modulation and coding, the multiple-access technique, and the MIMO transceiver techniques. The MAC layer determines how the PHY-layer resources are accessed. In packet-switched networks, the base station allocates the resources using a scheduler (Section 5.5.4) for both uplink and downlink transmission.

Base station design

The location of the base stations and the design of each base are critical for providing adequate coverage and capacity in the network. The base design includes the number of sectors per site and the antenna array architecture for each sector. The array consists of multiple antenna elements, and each element can be characterized by its beamwidth (horizontal and vertical) and the polarization (vertical, slant, cross-polarized). The physical dimensions of the antenna element are based on these characteristics and related to the carrier frequency as described in Section 5.5.1. The array is characterized by the positions and orientation of the antenna elements. For example, elements could be arranged in a linear configuration with close (half-wavelength) spacing or diversity (many wavelengths) spacing, as described in Section 5.5.1. The downtilt angle, the vertical position, and the azimuthal orientation of the array elements could be adjusted to affect the coverage.

The location and height of base station antennas are often constrained by zoning restrictions and must conform to local aesthetic values. These restrictions become more challenging as the number of antennas increases and the size of the array grows. Larger arrays also require more robust infrastructure to compensate for the additional weight and windloading effects.

The bases in a cellular network are connected to the core network through backhaul links. The core network provides access to the Internet or conventional phone network. The backhaul links are often wired connections running over copper wire or fiber optic cable, but they could also be implemented wirelessly on spectral resources orthogonal to those used for serving mobile terminals.

Terminal design

The mobile terminal consists of the signal processing capabilities and the antenna configuration. Most mobile handsets support one or two antennas. Larger mobile devices such as tablets or laptops have more real estate and computational power to support more antennas and more sophisticated algorithms.

Starting with a blank slate, an engineer would design the entire system—PHY and MAC layers, base stations, and terminals—from scratch. For a mature network where the base station infrastructure is already deployed, however, an engineer would have limited means of optimizing performance. Air interface parameters could be altered through software, and the downtilt and orientation of base station antennas could be adjusted to optimize coverage.

The context of the MIMO design strategies discussed in this and the following chapter lies between these two extremes. We are interested in evaluating MIMO techniques within the framework of an existing packet-based cellular network standard. Therefore, with regard to Figure 5.1, we assume the MAC and PHY layers are fully defined for SISO links, and we focus on the impact of the MIMO techniques and the antenna array architecture at the base and mobile. We assume that a generic OFDMA air interface partitions the bandwidth into parallel subbands so we can study the performance of MIMO in a narrowband channel.

5.2 System model

In practice, the location of base stations is irregular, and dependent on factors such as the terrain, zoning restrictions and user distribution. However, for the purposes of idealized performance evaluation, it is convenient to assume a regular placement of bases as shown in Figure 5.2, where the geographic area is partitioned into a hexagonal grid of cells and where a base is at the center of each cell site. As a result of different transmission powers and signal fading, the base closest to a user is not necessarily the one assigned to it. Therefore a user lying within a given cell is not necessarily assigned to that base, and the cell boundaries simply highlight the placement of the bases.

Each site is partitioned into S sectors, and a base station is associated with each sector. Each base is equipped with M antennas and serves K users, each with N antennas. The uplink received signal at base b $(b = 1, \ldots, B)$ is

$$\mathbf{x}_b = \sum_{k=1}^{KB} \mathbf{H}_{k,b}\mathbf{s}_k + \mathbf{n}_b \tag{5.1}$$

$$= \sum_{j \in U_b} \mathbf{H}_{j,b} \mathbf{s}_j + \underbrace{\sum_{j \notin U_b} \mathbf{H}_{j,b} \mathbf{s}_j}_{\text{interference}} + \mathbf{n}_b, \tag{5.2}$$

where

- $\mathbf{H}_{k,b} \in \mathbb{C}^{M \times N}$ is the block-fading MIMO channel between the kth user $(k = 1, \ldots, KB)$ and the bth base
- $\mathbf{s}_k \in \mathbb{C}^N$ is the transmitted signal from user k
- $\mathbf{n}_b \in \mathbb{C}^M$ represents receiver thermal noise with distribution $\mathcal{CN}(\mathbf{0}_M, \sigma^2 \mathbf{I}_M)$

Equation (5.2) separates the received signal in terms of desired signals from users assigned to base b (denoted by the set U_b) and interference signals from all other users. The transmit covariance for user k is $\mathbf{Q}_k := \mathbb{E}(\mathbf{s}_k \mathbf{s}_k^H)$, with a power constraint $\mathrm{tr}(\mathbf{Q}_k) \leq P_k$. (As we will describe in Section 5.2.2, this is actually a *scaled* power constraint that does not correspond to the actual maximum allowable radiated power.)

The downlink signal received by the kth user $(k = 1, \ldots, KB)$ is

$$\mathbf{x}_k = \sum_{b=1}^{B} \mathbf{H}_{k,b}^H \mathbf{s}_b + \mathbf{n}_k \tag{5.3}$$

$$= \mathbf{H}_{k,b^*}^H \mathbf{s}_{b^*} + \underbrace{\sum_{b \neq b^*} \mathbf{H}_{k,b}^H \mathbf{s}_b}_{\text{interference}} + \mathbf{n}_k, \tag{5.4}$$

where

- $\mathbf{H}_{k,b}^H \in \mathbb{C}^{N \times M}$ is the block-fading MIMO channel between the kth user and the bth base $(b = 1, \ldots, B)$
- $\mathbf{s}_b \in \mathbb{C}^M$ is the transmitted signal from base b
- $\mathbf{n}_k \in \mathbb{C}^N$ is the thermal noise vector with distribution $\mathcal{CN}(\mathbf{0}_N, \sigma^2 \mathbf{I}_N)$

Under the assumption that user k is assigned to base b^*, equation (5.4) separates the received signal in terms of the desired and interfering signals. The transmit covariance for base b is $\mathbf{Q}_b := \mathbb{E}(\mathbf{s}_b \mathbf{s}_b^H)$, with a (scaled) power constraint $\mathrm{tr}(\mathbf{Q}_b) \leq P$.

For both the uplink and downlink, we assume the components of the channel $\mathbf{H}_{k,b}$ are i.i.d. Rayleigh with distribution $h_{k,b}^{(m,n)} \sim \mathcal{CN}(0, \alpha_{k,b}^2)$. The variance $\alpha_{k,b}^2$ can be interpreted as the average channel gain between user k and base b relative to a user located at a reference distance d_{ref} from the base. This value depends on the distance-based pathloss, shadow fading realization, and direction between user k and base b:

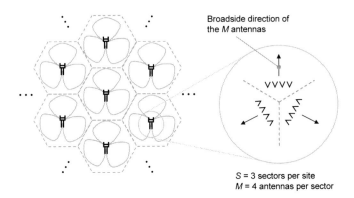

Fig. 5.2 Network of hexagonal cells. Each cell site is partitioned radially into S sectors, and each sector uses M antennas to serve K users.

$$\alpha_{k,b}^2 = \left(\frac{d_{k,b}}{d_{\mathrm{ref}}} \right)^{-\gamma} Z_{k,b} G_{k,b}, \tag{5.5}$$

where

- $d_{k,b}$ is the distance between user k and base b
- d_{ref} is the reference distance with respect to which the pathloss is measured, described in Section 5.2.2
- γ is the pathloss exponent
- $Z_{k,b}$ is the shadowing realization between user k and base b caused by large-scale obstructions such as terrain and buildings
- $G_{k,b}$ is the direction-based antenna response of the base antenna, described in the following section

The pathloss exponent value is $\gamma = 2$ for free space, but it is in the range of 3 to 4 for typical cellular channels. In the spatial channel model often used in 3GPP cellular standards, $\gamma = 3.5$ for suburban and urban macrocells with a carrier frequency of 1.9 GHz. The γ value was derived from a modified Hata urban propagation model which is applicable over a frequency range from 1.5 to 2.0 GHz [36]. The *next-generation mobile networking* (NGMN) simulation methodology [135] assumes that $\gamma = 3.76$ for a carrier frequency of 2.0 GHz.

In macrocellular outdoor environments, the shadowing realization is typically modeled with a log-normal distribution with standard deviation 8 dB:

$Z = 10^{(x/10)}$, $x \sim \eta(0, 8^2)$. The shadowing realization could be correlated between users (or bases) located near each other if they are similarly affected by common obstructions.

5.2.1 Sectorization

To implement sectorization, base antennas have a directional response that varies as a function of the angular (azimuthal) direction. The response $G_{k,b}$ for a directional antenna element is a function of the angular difference between the pointing (broadside) direction of the antenna and the direction of the user with respect to this element. As shown in Figure 5.3, ω_b is the broadside direction of sector antenna b, and $\theta_{k,b}$ is the direction of user k with respect to the location of sector b. The angles are measured with respect to a common reference (the positive x-axis). We will assume that multiple antennas belonging to a given sector have the same pointing direction and antenna response, but this assumption is not necessary in general.

When there is no sectorization, the bases employ omni-directional base antennas whose response is unity independent of the user location: $G_{k,b} = 1$. Under sectorization, we assume that each cell site is divided radially into S sectors. The broadside directions of the sector antennas is shown in Figure 5.4. For $S = 3$, the broadside directions of the three sectors are 90, 210, and 330 degrees, giving the so-called "cloverleaf" pattern of sector responses. For $S = 6$ and 12 sectors, the broadside direction of the sth sector ($s = 1, \ldots, S$) is $\frac{360(s-1)}{S}$ degrees.

Under *ideal* sectorization, the antenna gain is unity within the sector and zero outside:

$$G_{k,b}(\theta_{k,b} - \omega_b) = \begin{cases} 1 \text{ if } |\theta_{k,b} - \omega_b| \leq \frac{\pi}{S} \\ 0 \text{ if } |\theta_{k,b} - \omega_b| > \frac{\pi}{S}. \end{cases} \tag{5.6}$$

In normalizing the antenna gain to be unity within the sector, we implicitly assume that the total transmit power per site is fixed and that the reduction in power per sector is offset by the gain achieved with the narrower beam pattern. In other words, with S sectors per site, each sector transmits with a fraction $1/S$ and achieves a directional gain of factor S. These factors cancel one another so the gain is unity, independent of S.

In practice, there is intersector interference as a result of non-ideal antenna patterns. The response used in simulations is often a parabolic model that closely matches the response of commercial sector antennas:

$$G_{k,b}(\theta_{k,b} - \omega_b)(\text{dB}) = \max \left\{ -12 \left[\frac{(\theta_{k,b} - \omega_b)}{\Theta_{3 \text{ dB}}} \right]^2, A_s \right\}, \tag{5.7}$$

where $\Theta_{3 \text{ dB}}$ is the 3 dB beamwidth of the response and $A_s < 0$ is the sidelobe level of the response, measured in dB. This parabolic response is sometimes known as the Spatial Channel Model (SCM) sector response [36]. Note that the response $G_{k,b}(\theta_{k,b} - \omega_b) = -3$ dB when the user angle is half the 3 dB beamwidth: $\theta_{k,b} - \omega_b = \Theta_{3 \text{ dB}}/2$. As with the ideal sector response, the response in (5.7) is normalized under the assumption of fixed transmit power per site. The sector responses for ideal, parabolic, and commercially available antennas are shown in Figure 5.6 for the case of $S = 3$ sectors per site. The parabolic response, with $\Theta_{3 \text{ dB}} = 70\pi/180$ and $A_s = -20$ dB, is a good approximation of the commercial response over the main lobe and overestimates the sidelobe level. Therefore the simulated performance will overestimate the interference and result in a lower bound on capacity performance.

The 3 dB-beamwidth and sidelobe level parameters are dependent on the number of sectors per cell S. To ensure reasonable intersector interference, the beamwidth and sidelobe levels should decrease as the number of sectors per cell increases. The parameters we use in our simulations are summarized in Figure 5.5. We will refer to a site with more than $S = 3$ sectors as having *higher-order sectorization*. In going from $S = 3$ to 6 sectors, the beamwidth is reduced by a factor of two, and the antenna gain increases by 3 dB. However, because we fix the total transmit power per site, the power for each of the $S = 6$ sectors is halved. The antenna gain and power reduction cancel each other, so the response in the broadside direction is unity, and the sidelobe levels drop by 3 dB. The same scaling occurs in going from $S = 6$ to $S = 12$ sectors. The responses for $S = 3, 6, 12$ sectors per site are shown in Figure 5.7 for the parameters in Figure 5.5. We emphasize that these parameters are chosen to model well-designed sectorized antennas. Antenna responses of actual sectorized antennas may differ from this parabolic approximation.

The parabolic response is an approximation of an actual antenna's response measured in an anechoic chamber with zero angle spread. Because of the non-zero angle spread observed in practice, signals transmitted from the base will experience angular dispersion, resulting in a wider antenna response. The resulting antenna response can be obtained by convolving (in the angular domain) the channel's power azimuth spectrum (PAS) with the zero-angle-spread response in (5.7) [137]. The power azimuth spectrum of a typical urban macrocellular channel can be modeled as a Cauchy-Lorenz distribution with an RMS angle spread of $\phi = 8\pi/180$ radians [138]. The relatively narrow angle spread is due to the fact that the base station antennas are higher than the surrounding scatterers. Figure 5.8 shows the resulting antenna responses for $S = 3, 6, 12$ sectors. Because the PAS is relatively narrow compared to the baseline $S = 3$ response in Figure 5.7, the resulting convolved response is similar to the baseline response. However, the PAS is wide compared to the $S = 12$ baseline response, and the resulting convolved response is significantly wider than the baseline response.

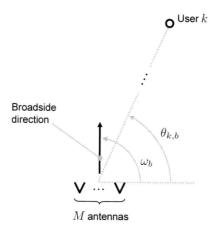

Fig. 5.3 The direction of user k with respect to the M antennas of sector b is given by $\theta_{k,b}$. The broadside direction of the M antennas is ω_b. These directions are measured with respect to the positive x-axis.

5.2.2 Reference SNR

The *reference SNR* is a single parameter that captures the effects of link budget parameters such as transmit power, antenna gain, cable loss, receiver noise, additive noise power spectral density, and bandwidth. Encapsulating these parameters is very useful because it allows us to present simulation results using a single parameter instead of a list of parameters.

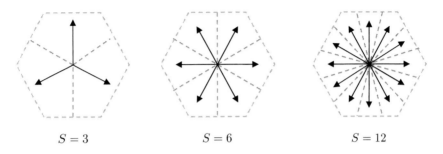

$S = 3$ $S = 6$ $S = 12$

Fig. 5.4 Broadside direction of sectorized antennas for $S = 3, 6, 12$ sectors per cell site.

The *downlink* reference SNR is the average SNR of a signal transmitted from a reference base and measured by a user at a reference location, as shown in Figure 5.9. The following assumptions are made in defining the reference SNR.

- The base transmits with full power P
- The user is is located at the reference location at the cell border with distance $d = d_{\text{ref}}$ from the reference base
- The user is in the boresight direction of the base so that the directional antenna response is $G = 1$
- The shadow fading realization is $Z = 1$

Under these conditions, the average channel gain (5.5) is $\alpha^2 = 1$. The received signal is $x = hs + n$, and the average SNR is

$$\frac{\mathbb{E}\left[|h|^2\right]\mathbb{E}\left[|s|^2\right]}{\mathbb{E}\left[|n|^2\right]} = \frac{\alpha^2 P}{\sigma^2} = \frac{P}{\sigma^2}. \tag{5.8}$$

Therefore the downlink reference SNR is simply $\frac{P}{\sigma^2}$. Because the SNR is measured at the cell edge, the reference SNR is sometimes known as the cell-edge SNR. While we have considered the SISO channel here, the reference SNR does not depend on the number of transmit antennas if the channel is i.i.d. and if the total transmit power is constrained to be P.

For the uplink, we assume that the total power constraint of the K users assigned to a base is P and that the power constraint for each user is the same: $P_k = P/K$, for $k = 1, \ldots, K$. We define the uplink reference SNR as

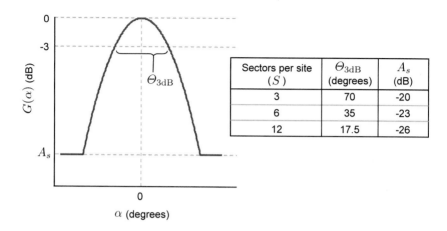

Fig. 5.5 Parabolic sector response (5.7). The response is parameterized by the 3 dB beamwidth (Θ_{3dB}) and sidelobe level (A_s).

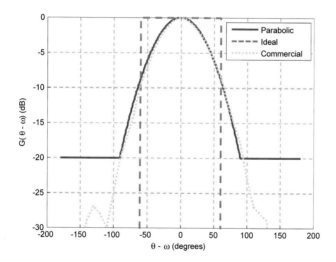

Fig. 5.6 Sector responses for $S = 3$ sectors per site. The parabolic response is from (5.7), and the ideal response is given by (5.6). The parabolic response matches the response of the commercially available antenna over the main lobe and overestimates the sidelobe level.

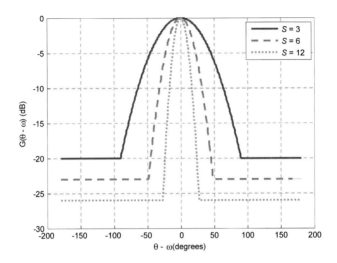

Fig. 5.7 Parabolic responses for $S = 3, 6, 12$ sectors per site, given by (5.7) and parameters in Figure 5.5.

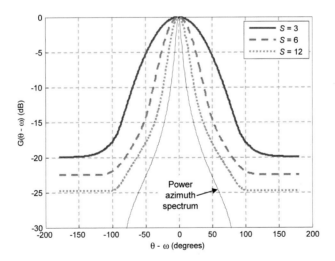

Fig. 5.8 Sector responses for $S = 3, 6, 12$ sectors per site obtained by convolving the power azimuth spectrum (for 8 degree angle spread) with the parabolic response in Figure 5.7.

the average total SNR received by the reference base when the K users at
the reference location transmit with full power: $\frac{P}{\sigma^2}$.

Recall that for the SU- and MU-MIMO channels, the SNR is similarly
defined as $\frac{P}{\sigma^2}$. If the isolated channels are normalized to have unity gain
$\mathbb{E}\left[|h|^2_{i,j}\right] = 1$, then P is both the transmit power constraint and the average
received power. For the downlink cellular channel, we see now that P is the
power constraint scaled with respect to the reference location. The actual
radiated power constraint can be determined from P by accounting for the
channel loss from the reference base to reference user location.

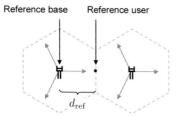

Fig. 5.9 The reference SNR P/σ^2 is measured by a user located in the boresight direction
of the serving base at distance d_{ref}, which is half the intersite distance.

The reference SNR depends on the link budget parameters, and is usually
different for the uplink and downlink. Examples of reference SNR values and
their associated parameters are given in Figure 5.10. The parameters for the
two downlink examples have been chosen to give respective guidelines for
high and low reference SNRs. *Downlink A* is an example of a base station
transmitter with relatively high power (40 watts) and its parameters correspond to the NGMN model. The reference received signal power and noise
power are respectively

$$P = 46 \text{ dBm} - 9 \text{ dB} + 14 \text{ dB} - 128 \text{ dB} + 37.6\log_{10}\left(\frac{1000}{250}\right) \text{dB} = -54.3 \text{ dBm}$$

$$\sigma^2 = -174 \text{ dBm} + 10\log_{10}(10^7) \text{ dB} = -104 \text{ dBm},$$

and the reference SNR is $P/\sigma^2 = 49.6$ dB. The NGMN downlink simulation
parameters are similar to those given by Downlink A, except that there is a
20 dB in-building loss, resulting in a cell-edge SNR of 29.6 dB. *Downlink B* is
a base station with lower transmit power and larger inter-site distance. With

an intersite distance of 2000 m, each cell has an area of about 3 km^2, and covering a large city with an area of 1000 km^2 would require several hundred bases. Downlink B also includes a 20 dB in-building penetration loss, resulting in a reference SNR of 1.4 dB. Because of the significant attenuation of signals passing through walls, indoor coverage is achieved more effectively by using indoor bases.

Parameters for Uplink A and B give respective upper and lower guidelines for the uplink reference SNRs. Except for transmit powers and receiver noise figures, the parameters are the same for both uplink and downlink. Typically, transmit powers are higher for the downlink because base stations have larger power amplifiers, and receiver noise is also higher in the downlink due to noisier electronics at the handset. The NGMN uplink parameters are similar to Uplink B, except that the reference distance is 250 m, resulting in a reference SNR of 11.6 dB.

The pathloss intercept has a logarithmic dependency on the carrier frequency f (measured in units of mega-Hertz) given approximately by $B \log_{10} f$, where the parameter B depends on parameters such as the antenna height and terrain type. Its value is approximately 30 for typical macrocellular deployments [36] so a doubling of the carrier frequency from 1000 MHz to 2000 MHz would result in an additional pathloss of about 10 dB. Therefore to maintain the same coverage for higher carrier frequencies, higher transmit power or higher antennas are required.

5.2.3 Cell wraparound

For a given downlink cellular network, users at the edge of the network will experience less interference than those at the center. In order to make the interference statistics the same regardless of user location, we can approximate an infinite cellular network by projecting a finite network on a torus. Cells at one edge of the network are "wrapped around" the torus to be adjacent to cells at the opposite edge. A wrapping strategy based on a hexagonal network of hexagonal cells has desirable characteristics because it exhibits translational and rotational invariance [139].

Figure 5.11 shows a hexagonal wrapping using a network of $B = 7$ cells labeled 1a, 2a,..., 7a. This network is replicated six times around the perimeter. The neighbor to the northwest of cell 6a is cell 4e, which is a replica

	Downlink A	Downlink B	Uplink A	Uplink B
Transmit power	46 dBm (40 W)	38.6 dBm (7.2 W)	30 dBm (1 W)	24 dBm (250 mW)
Noise power spectral density (N_0)	-174 dBm/Hz	-174 dBm/Hz	-174 dBm/Hz	-174 dBm/Hz
Bandwidth (W)	10 MHz	10 MHz	10 MHz	10MHz
Receiver noise figure	9 dB	9 dB	5 dB	5 dB
Net antenna gain	14 dB	14 dB	14 dB	14 dB
In-building penetration loss	0 dB	20 dB	0 dB	0 dB
Pathloss intercept, 1000m	128 dB	128 dB	128dB	128 dB
Pathloss exponent (γ)	3.76	3.76	3.76	3.76
Intersite distance ($2d_{\text{ref}}$)	500 m	2000 m	500 m	2000 m
Reference distance (d_{ref})	250 m	1000 m	250 m	1000 m
Reference SNR	**49.6 dB**	**0.0 dB**	**37.6 dB**	**-11.0 dB**

Fig. 5.10 Examples of link parameters and corresponding reference SNRs. Downlink A and Uplink A are relatively high reference SNRs because they use high-powered transmitters with small intersite distances. Downlink B and Uplink B are relatively low reference SNRs because they user low-powered transmitters with large intersite distances and a significant in-building penetration loss.

of cell 4a. In the simulations, users are placed with a uniform distribution over the original network of B cells. Using cell wraparound, the channel gain $\alpha_{k,b}^2$ between a user k and a given base $b = 1, \ldots, B$ is the maximum of the channel gains between the user and the seven wraparound replicas of base b, where the shadowing realization for all replicas is the same. For example, the channel gain between user k and cell 1 is

$$\alpha_{k,1}^2 = \max \left[\alpha_{k,1a}^2, \alpha_{k,1b}^2, \ldots, \alpha_{k,1g}^2 \right] \tag{5.9}$$

$$= \max \left[\left(\frac{d_{k,1a}}{d_{\text{ref}}} \right)^{-\gamma} Z_{k,1} G_{k,1a}, \ldots, \left(\frac{d_{k,1g}}{d_{\text{ref}}} \right)^{-\gamma} Z_{k,1} G_{k,1g} \right]. \tag{5.10}$$

Repeating this procedure for all B cells, we see that the channel variances for a given user placement are, in effect, measured with respect to the B cells with the minimum radio distance. For the case of omni-directional base antennas, $G_{k,1a} = \cdots = G_{k,1g} = 1$, and

$$\alpha_{k,1}^2 = \left[\frac{\min \left(d_{k,1a}, d_{k,1b}, \ldots, d_{k,1g} \right)}{d_{\text{ref}}} \right]^{-\gamma} Z_{k,1}. \tag{5.11}$$

In this case, the channel gains are measured with respect to the B cells with the minimum *Euclidean* distance. For example in Figure 5.11, the B channel gains for a user placed in cell 6a are with respect to the cells highlighted by the black border. For cell wraparound to be effective, the size of the network should be sufficiently large that the channel gain of the dominant replica in (5.9) is several orders of magnitude larger than the others. For a pathloss coefficient of $\gamma \approx 4$, a network of $B = 19$ cells is sufficiently large. (We used a smaller network with $B = 7$ cells just for illustrative purposes.) While we have described the wrapping technique for the downlink, it can be applied to the uplink in a similar fashion.

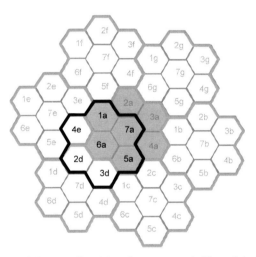

Fig. 5.11 Network of $B = 7$ cells with cell wraparound. The original cells (1a, 2a,...,7a) are shaded gray. The B channel gains for each user are measured with respect to the network of B cells with the smallest radio distance. With omni-directional base antennas, the smallest radio distance becomes the smallest Euclidean distance. For example, a user placed in cell 6a measures its channel gain with respect to the B cells highlighted by the black border.

5.3 Modes of base station operation

In conventional cellular networks, communication within each cell is independent of the other cells. Interference caused to other cells is treated as noise, and if the interference is stronger than the thermal noise, the performance becomes interference limited. As mentioned in Chapter 1, interference can be mitigated in the time and frequency domains. In this section, we describe how interference can be mitigated in the *spatial* domain using the MIMO techniques described in previous chapters. The efficiency of the mitigation depends on the mode of base station operation. Base stations can operate independently or in a coordinated fashion. These options are highlighted in Figure 5.12 and described below.

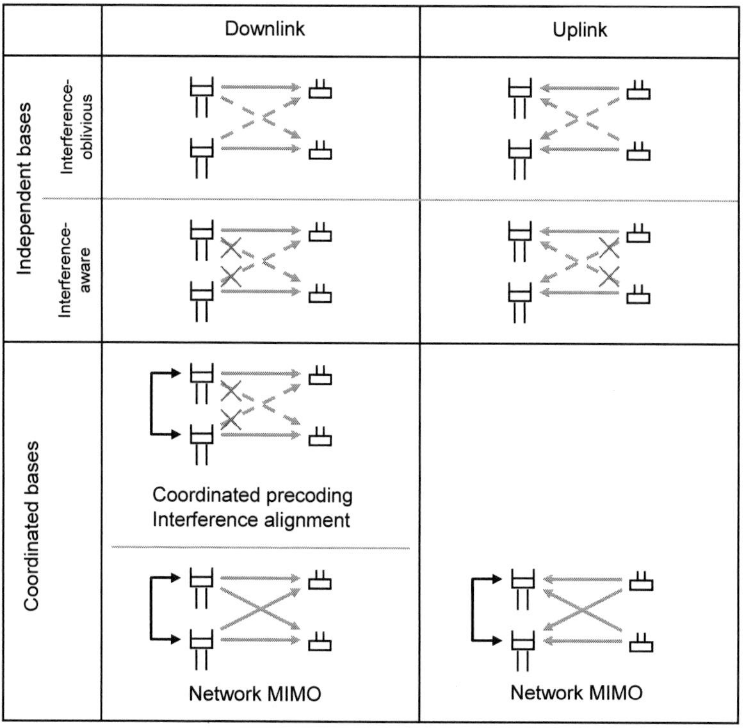

Fig. 5.12 Modes of operation for cellular networks. Solid lines indicate signals to assigned users (downlink) or from assigned users (uplink). Dashed lines indicate signals that cause interference. An "x" over a dashed line indicates significantly mitigated interference. The double arrows indicate coordination between bases.

5.3.1 Independent bases

Under independent base station operation, no information is shared between the bases. Interference in the spatial domain can be mitigated in an *interference-oblivious* or *interference-aware* manner.

Interference-oblivious mitigation

Interference-oblivious mitigation refers to spatial processing performed at a transmitter or receiver which does not account for the channel state information (CSI) of, respectively, unintended receivers or interfering transmitters.

In an uplink SIMO network where each user has a single antenna and where the base has M antennas, interference-oblivious mitigation is implemented with a matched filter (maximal ratio combining) receiver. Let us consider an uplink SIMO channel (5.2) with the received signal at the assigned base given by

$$\mathbf{x} = \mathbf{h}_k s_k + \sum_{j \neq k} \mathbf{h}_j s_j + \mathbf{n}, \tag{5.12}$$

where k is the index of the desired user and where the signals from all other users $j \neq k$ cause interference. The matched filter receiver output $\mathbf{h}_k^H \mathbf{x}$ has an SINR given by (2.58), and the achievable rate is

$$R = \log_2 \left(1 + \frac{\|\mathbf{h}_k\|^4 P}{\|\mathbf{h}_k\|^2 \sigma^2 + \sum_{j \neq k} \left|\mathbf{h}_k^H \mathbf{h}_j\right|^2 P_j} \right). \tag{5.13}$$

If there is no interference ($P_j = 0$ for all $j \neq k$), then the signal is corrupted by spatially white noise, and the matched filter is optimal (Section 2.3.1). Due to the presence of intercell interference, the matched filter is no longer optimal. However, even though this receiver has no knowledge of the interferers' CSI ($\mathbf{h}_j, j \neq k$), interference can be mitigated if a significant fraction of the power lies in the null space of the desired channel $\mathbf{h}_k$. For example, interference is totally mitigated if interference channels are orthogonal to the desired channel: $\mathbf{h}_k^H \mathbf{h}_j = 0$, for all $j \neq k$.

For a given channel $\mathbf{h}_k$, the statistics of the rate (5.13) depend on the relationship between $\mathbf{h}_k$ and the distribution of the interferers' channels ($\mathbf{h}_j, j \neq k$). For the case of i.i.d. Rayleigh channels, the average rate can be written as:

$$\mathbb{E}_{\mathbf{h}_j}(R) \geq \log_2\left(1 + \frac{\|\mathbf{h}_k\|^4 P}{\|\mathbf{h}_k\|^2 \sigma^2 + \sum_{j\neq k} \mathbb{E}_{\mathbf{h}_j}\left|\mathbf{h}_k^H \mathbf{h}_j\right|^2 P_j}\right) \tag{5.14}$$

$$= \log_2\left(1 + \frac{\|\mathbf{h}_k\|^4 P}{\|\mathbf{h}_k\|^2 \sigma^2 + \sum_{j\neq k} \|\mathbf{h}_k\|^2 \alpha_j^2 P_j}\right) \tag{5.15}$$

$$= \log_2\left(1 + \|\mathbf{h}_k\|^2 \frac{P_k}{\sigma^2 + \sum_{j\neq k} \alpha_j^2 P_j}\right), \tag{5.16}$$

where (5.14) follows from Jensen's inequality applied to the convex function $f(\mathbf{a}) = \log_2(1 + (1 + \mathbf{a}^H \mathbf{Ra})^{-1})$ and where (5.15) follows from the assumption of i.i.d. Rayleigh channels. The expression in (5.16) corresponds to the capacity of a SIMO channel $\mathbf{h}_k$ where the interference is spatially white with covariance $\mathbf{I}_N(\sigma^2 + \sum_{j\neq k} \alpha_j^2 P_j)$. Therefore by using a matched filter, the average rate achievable in the presence of colored interference is at least as high as the rate achievable in the presence of spatially white interference with the same power.

For the downlink, interference-oblivious mitigation for MISO channels can be implemented using maximal ratio transmission (Section 1.1.2.3) if the base station has knowledge of the desired user's channel. Similar to the uplink, interference can be implicitly mitigated if the channels to the unintended users are not aligned with the desired user's channel.

Interference-aware mitigation

Under interference-aware mitigation at the receiver, the spatial processing uses explicit channel knowledge of the interfering transmitters. For example, for the uplink SIMO channel (5.12), interference-aware mitigation for the desired user k can be implemented with an MMSE receiver (Section 2.3.1) using knowledge of $\mathbf{h}_k$ and the interfering users' channels $\mathbf{h}_j, j \neq k$. The receiver does not make any distinction between interfering users assigned to the base serving user k or assigned to other bases. CSI estimates could be obtained by demodulating pilot signals sent by each user. Since each user would transmit a pilot signal for coherent demodulation by its assigned base, interference-aware mitigation at the receiver can be achieved with no additional signaling overhead.

Because the MMSE receiver uses explicit knowledge of the interferers' channels, the MMSE receiver is more efficient than the matched filter for mitigating interference. Using the expression for the SINR at the output of the MMSE receiver (2.61), the achievable rate is

$$R = \log_2 \left[1 + \frac{P_k}{\sigma^2} \mathbf{h}_k \left(\mathbf{I}_M + \sum_{j \neq k} \frac{P_j}{\sigma^2} \mathbf{h}_j \mathbf{h}_j^H \right)^{-1} \mathbf{h}_k \right]. \qquad (5.17)$$

Because the MMSE receiver maximizes the SINR among all linear receivers, its output SINR will be greater than that of the matched filter, and its rate (5.17) will be greater than that of the matched filter (5.13) for any set of channel realizations. It follows that for a fixed channel $\mathbf{h}_k$, the MMSE rate (5.17) averaged over the interfering channels will be greater than the lower bound for the average MF rate (5.16).

Interference-aware mitigation can also be implemented for MIMO channels using a pre-whitening filter. (We recall that the MMSE receiver for a multiuser SIMO channel is equivalent to pre-whitening the colored interference followed by matched filtering.) We consider an uplink MIMO channel (5.2) where single-user MIMO is implemented. The received signal at the assigned base for user k is

$$\mathbf{x} = \mathbf{H}_k \mathbf{s}_k + \sum_{j \neq k} \mathbf{H}_j \mathbf{s}_j + \mathbf{n}, \qquad (5.18)$$

where the channels $\mathbf{H}_j, j \neq k$ are for the interfering users assigned to other bases. The transmit covariances for these users $\mathbf{Q}_j = \mathbb{E}(\mathbf{s_j} \mathbf{s}_j^H), j \neq k$ are set according to the channels of their respective serving bases. Assuming the interference covariance

$$\mathbf{Q}_I := \sigma^2 \mathbf{I}_M + \sum_{j \neq k} \mathbf{H}_j \mathbf{Q}_j \mathbf{H}^H \qquad (5.19)$$

is known by the base serving user k, the output of the pre-whitening filter is $\mathbf{Q}_I^{-1/2} \mathbf{x}$. If the transmit covariance for the desired user is $\mathbf{Q}_k$, the achievable rate is

$$R = \log_2 \det \left[\mathbf{I}_M + \mathbf{Q}_I^{-1} \mathbf{H}_k \mathbf{Q}_k \mathbf{H}_k^H \right]. \qquad (5.20)$$

In a similar manner, pre-whitening can be applied at the mobile receiver for downlink single-user MIMO transmission, and more generally, pre-whitening can also be applied for multiuser MIMO channels. For example, given the uplink received signal at base b (5.2), the pre-whitening is performed with respect to the interference covariance

$$\sigma^2 \mathbf{I}_M + \sum_{j \notin U_b} \mathbf{H}_{j,b} \mathbf{Q}_j \mathbf{H}_{j,b}^H. \qquad (5.21)$$

On the downlink, given the received signal by user k (5.4), the pre-whitening is performed with respect to the interference covariance

$$\sigma^2 \mathbf{I}_N + \sum_{b \neq b^*} \mathbf{H}_{k,b}^{H} \mathbf{Q}_b \mathbf{H}_{k,b}. \tag{5.22}$$

Interference-aware mitigation can be implemented at the transmitter if the channels of the unintended receivers are known. However, acquiring reliable channel estimates may be difficult in practice, especially for downlink transmission in FDD systems (Section 5.5.4). Linear precoding techniques such as those described in Section 4.2.1 could be used. In general, more sophisticated non-linear techniques could be used for interference-aware mitigation at the transmitter or receiver.

5.3.2 Coordinated bases

Compared to using independent bases, interference can be mitigated to a greater extent by coordinating the transmission and reception among groups of spatially distributed base stations. Coordination requires backhaul between bases to have higher bandwidth in order to support the sharing of information among them. Depending on the coordination technique, this information could consist of CSI, baseband signals, and/or user data bits. The bottom half of Figure 5.12 summarizes the coordination techniques which are described below.

Coordinated precoding

When bases operate independently, the precoding at a given base is a function of the *local* CSIT, which refers to the channels between the base and the surrounding users. With coordinated precoding, CSIT is shared between bases so that the precoding at a given base could be a function of CSIT as seen by multiple bases. (The implementation of CSIT sharing is described below in Section 5.5.5.) Most generally, the precoding at each base could be a function of the *global* CSIT, meaning the CSIT between all bases and all users. The transmitted signal at each base requires global CSIT but only local data, meaning each base transmits data only to users assigned to it. Global CSIT allows the precoding for all the bases to be computed jointly to optimize the performance criterion.

A special case of coordinated precoding for the case of directional beams is *coordinated beam scheduling*. As shown in Figure 5.13, global CSIT would allow each base to realize the potentially hazardous situation on the left where co-located users are assigned to different bases but served on the same time-frequency resources. To avoid this situation which would result in high interference for all users, the bases would schedule the co-located users on different resources.

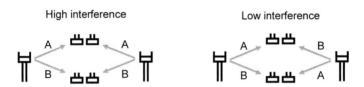

Fig. 5.13 On the left, users in each pair of co-located users are served by different bases but on the same time-frequency resource, resulting in high interference. Using coordinated beam scheduling, the bases would perform joint scheduling so that co-located users are served on different resources.

Interference alignment

Under interference alignment, precoding is implemented at each base so that the interference at each terminal is aligned to lie in a common subspace with minimum dimension, and the remaining dimensions can be used for interference-free communication with the assigned base. Like coordinated precoding, interference alignment requires global CSIT for precoding at each base. However unlike coordinated precoding, which can be implemented on a frame-by-frame basis, precoding for interference alignment occurs over an infinite number of independently faded time or frequency realizations [140].

As an example, we consider an interference channel with three single-antenna bases, each communicating with a single-antenna user as shown in Figure 5.14. Each user attempts to decode the signal from its assigned base in the presence of interference from the other two bases. Communication occurs over F time frames. We let $\mathbf{H}[f]$ represent where the 3×3 channel during frame $f = 1, \ldots, F$, and the channel is assumed to be spatially and temporally rich so the distribution is i.i.d. Rayleigh with respect to time and space. Interference alignment uses precoding (across time) at each transmitter such that as F goes to infinity, an average of $3/2$ streams per frame are received interference-free by the users. In general, for a system with K transmit-receive pairs (where each node has a single antenna), interference alignment asymptotically achieves $K/2$ interference-free streams per frame [140]. This $K/2$ factor is the multiplexing gain of the sum-capacity at high SNR. For comparison, under TDMA transmission each of K transmit/receive pairs would be active for a fraction $1/K$ of the time, achieving a multiplexing gain of only 1. If bases transmit simultaneously and independently, each user receives interference from the other bases, and the multiplexing gain is zero. If the channels among the K pairs were mutually orthogonal, a multiplexing gain of K could be achieved. Hence interference alignment enables noise-limited performance and achieves half the degrees of freedom of an interference-free channel.

In practice, interference alignment is difficult to implement because it requires very large time (or frequency) expansion and very rich channels. An alternative technique for static channels requires infinite resolution of channel realizations [141]. Techniques for achieving the "$K/2$" degrees of freedom in practical scenarios is an area of active research.

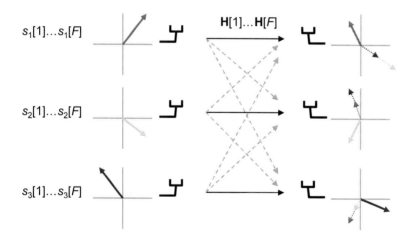

Fig. 5.14 Interference alignment for an interference channel with three transmit/receive pairs. Interference at each receiver is aligned to lie in a minimum-dimensional "garbage" subspace so the remaining dimensions can be used for interference-free communication with the intended transmitter. In this example, one stream per pair can be communicated interference-free.

Network MIMO

Under network MIMO, base stations operate in a fully coordinated manner as if they belonged to a single virtual base with distributed antennas. For example on the downlink, a coordination cluster of B base stations, each with M antennas, operates over a broadcast channel with BM transmit antennas [142]. The data for each user is in general transmitted from all bases in the cluster, and the precoding at each base depends on the global CSIT. Therefore downlink network MIMO requires both global CSIT and global data of all users at each base, resulting in higher backhaul bandwidth requirements than coordinated precoding and interference alignment. Downlink network MIMO also requires tight synchronization and phase coherence among coordinated bases so that signals transmitted from different bases arrive at each user with the proper phase.

Network MIMO can be implemented on the uplink in a similar fashion by using multiuser detection to jointly detect users across multiple bases [143]. A cluster of B coordinated bases, each with M antennas, implementing network MIMO could be viewed as a single virtual base with BM distributed receive antennas. One could therefore implement uplink network MIMO by collecting the baseband signals from a cluster of bases and performing joint detection (for example, MMSE-SIC) at a central location. For a system with multiple-antenna users, scale-optimal multiplexing gain can be achieved if the number of users exceeds BM by transmitting a single stream per user without precoding or by using only local CSIT at each mobile (Section 4.1.1). Therefore significant performance gains using uplink network MIMO can be achieved without global CSIT knowledge, making it much easier to implement than downlink network MIMO. As a lower-complexity alternative to joint detection, one could selectively cancel interference by exchanging decoded data of dominant interferers among cooperating cells and reconstructing the interference signals. This technique achieves similar performance gains to joint detection network MIMO for low-geometry users but with significantly reduced backhaul overhead [144].

Figure 5.15 shows different modes of coordination for downlink network MIMO. (These modes apply to the uplink as well.) With non-overlapping clusters, the network is partitioned into fixed coordination clusters. Each user is assigned to a single cluster and is served by the bases belonging to that cluster. In the left subfigure of Figure 5.15, the network is partitioned into two clusters of $B = 2$ bases each. Users will experience interference from adjacent clusters, and those at the edge of the cluster will experience more interference

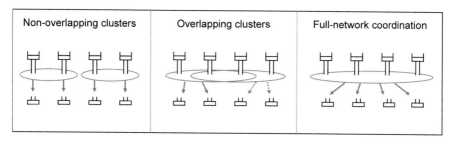

Fig. 5.15 Coordination modes for network MIMO. Interference occurs between clusters. Under full-network coordination, all bases in the network are coordinated and, under ideal conditions, performance is limited by thermal noise.

than those near the center. To reduce the intercluster interference, one could form user-specific clusters of B bases so that each user is in the center of its assigned cluster. In general, different clusters consist of different sets of bases, and a given base may belong to multiple clusters. Different clusters that consist of at least one common base are said to overlap, as illustrated in the center subfigure of Figure 5.15 where two clusters of $B = 3$ bases overlap and share two common bases. For a given cluster size B, the user-specific clusters are more general than the fixed clusters; hence the geometry statistics with user-specific, overlapping clusters could be better than those with fixed, non-overlapping clusters. Network MIMO with overlapping clusters is described in Section 5.4.4 for minimum sum-power transmission, and it is addressed in [145] for full-power transmission. A survey of techniques for network MIMO with clustering is given in [146].

With either non-overlapping or overlapping clusters, the system performance is limited by intercluster interference. However, if all the bases in the entire network are coordinated so that all users are served by a single coordination cluster, the network performance would be noise limited. Network-optimal performance could be achieved if capacity-achieving strategies (DPC on the downlink and MMSE-SIC on the uplink) are ideally implemented across the entire network [142]. A survey of analytical performance results for full network coordination in the context of the Wyner model can be found in [146]. Most of these results assume ideal coordination and perfect knowledge of global CSI. In practice, accurate CSIR is difficult to obtain, especially for larger networks where separation between transmitters and receivers is large.

For low-mobility users, the CSI quality is sufficiently good to support full network coordination for a small indoor network [147]. However, for larger networks, acquiring reliable global CSI and sharing the information across the network becomes a significant challenge.

Figure 5.16 summarizes the features of downlink MIMO options in the context of a network with B bases. Each base $b = 1, \ldots, B$ is equipped with M antennas and is assigned to an M-antenna user with index b. The data signal for user b is denoted as u_b. The channel between base b and user k is $\mathbf{H}_{k,b}^H$.

For interference-oblivious precoding with closed-loop SU-MIMO, base b requires knowledge of only user b and its channel $\mathbf{H}_{b,b}^H$. If the spectral resources are partitioned into B mutually orthogonal blocks and if each base transmits on one of these blocks, a multiplexing gain of M/B per base can be achieved. Under interference-aware precoding, base b requires local CSIT with respect to all users $\mathbf{H}_{k,b}^H, k = 1, \ldots, B$. Both coordinated precoding and interference alignment require global CSIT for each base b: $\mathbf{H}_{k,j}^H, j, k = 1, \ldots, B$. The multiplexing gain per base with interference alignment is $M/2$ [140] in the limit of an infinite time horizon. This gain is an upperbound for the gain achieved with either coordinated precoding or interference-aware precoding, both of which operate on a frame-by-frame basis. Using full coordination among the B bases, network MIMO requires both global CSIT and global data and achieves a multiplexing gain per base of M.

5.4 Simulation methodology

In this section, we describe the metrics and methodologies used for numerically evaluating the system-level performance of MIMO cellular networks. Simulations are performed over a large network of hexagonal cells with B bases and a total of KB users in the network. For a given network architecture, users are dropped randomly and channel realizations are generated for many iterations. For a given iteration, the channel realizations are fixed, and the performance for each base can be characterized by the K-dimensional capacity region $\mathcal{C}$ obtained by treating the interference in (5.2) and (5.4) as noise. To determine the optimal rate vector $\mathbf{R}^{\mathrm{Opt}} \in \mathcal{C}$, we use ei-

Base mode	Independent bases		Coordinated bases		
Technique	**Interf.-oblivious precoding**	**Interf.-aware precoding**	**Coordinated precoding**	**Interference alignment**	**Network MIMO precoding**
Description	Precoding at base b is a function of the desired user's channel	Precoding at base b is a function of the desired and unintended users' channels	Precoding at base b is a function of channels seen by all coordinated bases	Signals from interfering bases are aligned to lie in a "garbage" subspace	A cluster of coordinated bases transmit as a single base with distributed antennas
Benefits	No exchange of base information	Some gains for cell edge users	Some gains over interf.-aware precoding	Interference-free performance	Potentially optimal with full-network coordination
Drawbacks	No explicit interference mitigation.	Minimal improvement in throughput	Requires global CSIT	Requires global CSIT, high complexity.	Requires global CSIT and data, high complexity
Multiplexing gain per base	M/B	$\leq M/2$	$\leq M/2$	$M/2$	M
Base b knowledge	u_b $\mathbf{H}_{b,b}^{H}$	u_b $\mathbf{H}_{k,b}^{H}, k = 1,\dots,B$	u_b $\mathbf{H}_{k,j}^{H}, k,j = 1,\dots,B$		$u_k, k = 1,\dots,B$ $\mathbf{H}_{k,j}^{H}, k,j = 1,\dots,B$

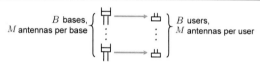

B bases, M antennas per base B users, M antennas per user

Fig. 5.16 Overview of spatial interference mitigation techniques for downlink cellular networks. Multiplexing gain per base is given for a system with B bases, each serving one user, and each base and user has M antennas.

ther *proportional-fair* or *equal-rate* performance criteria (described in Section 5.4.1) to achieve fairness among the K users.

These two methodologies will be used in the following chapter to evaluate system performance of different MIMO techniques. The proportional-fair methodology will be used primarily for evaluating the performance of independent bases, and the equal-rate methodology will be used for evaluating the performance of coordinated bases.

An iterative scheduling algorithm described in Section 5.4.2 can be used to numerically calculate the optimum rate vectors under either criterion, and it is also used in practice for scheduling in time-varying channels (Section 5.5.4). Rate vectors are collected over many iterations and across all bases in the network to generate throughput and user-rate metrics. Details of the proportional-fair and equal-rate simulation methodologies are given in Sections 5.4.3 and 5.4.4, respectively.

5.4.1 Performance criteria

In previous chapters, the sum-rate capacity is used for multiuser MIMO channels to determine the rate vector which maximizes the sum rate of the users. For a base station serving K users with capacity region $\mathcal{C}$, the sum-capacity rate vector is given by

$$\mathbf{R}^{\mathrm{Opt}} = \arg\max_{\mathbf{R}\in\mathcal{C}} \sum_{k\in U} R_k, \qquad (5.23)$$

where U is the set of K users assigned to this base. If the average SNRs of the users are the same, then this criterion is fair in the sense that users would receive the same average rate over many channel realizations. However, due to co-channel interference in cellular networks, the geometry of the users can vary significantly. The maximum sum-rate criterion would then not be fair because maximizing the sum rate favors service to users with high geometry while denying service to those with low geometry.

To provide more fairness, the *proportional-fair* criterion provides a higher incentive to serve low-geometry users than sum-rate maximization. This is achieved by maximizing the sum of log rates

$$\mathbf{R}^{\mathrm{Opt}} = \arg\max_{\mathbf{R}\in\mathcal{C}} \sum_{k} \log R_k \qquad (5.24)$$

subject to the transmitter power constraint. It can be shown that an equivalent condition for the proportional-fair vector $\mathbf{R}^{\mathrm{Opt}}$ in (5.24) is [148]

$$\sum_{k} \frac{R_k - R_k^*}{R_k^*} \le 0 \qquad (5.25)$$

for any $[R_1, \ldots, R_K] \in \mathcal{C}$. This means that in moving from the proportional-fair vector $\mathbf{R}^{\mathrm{Opt}}$ to any other rate vector $\mathbf{R} \in \mathcal{C}$, the sum of the fractional increases in user rates cannot be positive.

In the simulations for independent base operation in Chapter 6, each base on the downlink is assumed to transmit with full power P and, on the uplink, each user is assumed to transmit with full nominal power set according to a power control algorithm. If there is only a single user per base ($K = 1$), fairness is not a concern. In this case, each transmitter operates with full power and the optimization in (5.24) becomes trivial.

As an alternative to the proportional-fair criterion, one could provide fairness by maximizing the minimum rate achieved by any user:

$$\mathbf{R}^{\mathrm{Opt}} = \arg \max_{\mathbf{R} \in \mathcal{C}} \min_{k \in U} R_k, \qquad (5.26)$$

where U is the set of users served by the base. The solution provides *equal-rate* service to all users. The equal-rate criterion provides more fairness than the proportional-fair criterion because there is no difference among the users' rates. However, in cases with disparate SNRs, fairness could come at the expense of lower throughput because the system could be forced to allocate significant resources to the weakest user. As highlighted in Figure 5.17, the goal of maximizing base station throughput is at odds with providing fairness among users.

Performance criterion	Maximizes	Base station throughput	User fairness	Transmission mode
Maximum sum rate	Sum of rates	High	High	Full-power
Proportional fair	Sum of log rates	Medium	Medium	Full-power
Equal rate	Minimum rate	Low	Low	Minimum sum-power

Fig. 5.17 Criteria for evaluating cellular network performance. Numerical simulations in Chapter 6 will typically use the proportional-fair criterion for independent base operation and the equal-rate criterion for coordinated base operation.

Figure 5.18 shows the two-user rate regions for a time-division multiple-access (TDMA) channel, a multiple-access channel (with fixed covariances) and a broadcast channel. In the TDMA channel, the base station is restricted to serve one user at a time. Serving user 1 only, the rate vector $(R_1^*, 0)$ can be achieved. Serving user 2 only, the rate vector $(0, R_2^*)$ can be achieved. Other points on the rate region boundary can be achieved with time sharing. If $R_1^* > R_2^*$, the sum-rate-maximizing rate vector is $(R_1^*, 0)$. The equal-rate vector is the point (R_1, R_2) on the rate region boundary such that $R_1 = R_2$. The proportional-fair vector, $(R_1^*/2, R_2^*/2)$, can be shown to satisfy (5.25) and is achieved by serving each user half the time. It achieves a higher sum rate than the equal-rate vector and achieves better fairness than the maximum-sum-rate vector.

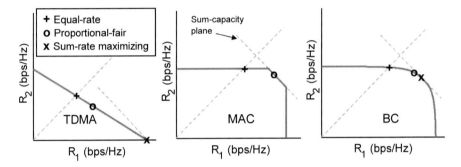

Fig. 5.18 2-user capacity regions for TDMA, MAC (with fixed covariances), and BC channels. Operating points for the three performance criteria are highlighted. For the MAC, any vector lying on the sum-capacity plane maximizes the sum rate.

5.4.2 Iterative scheduling algorithm

Given the rate region $\mathcal{C}$, we can find the sum-capacity rate vector (5.23) using the numerical optimization techniques described in Section 3.5. The proportional-fair rate vector (5.24) can be found by maximizing the sum of log rates $f(\mathbf{R})$ over $\mathcal{C}$. Because this is a standard convex program, convex optimization algorithms based on the ones in Section 3.5 can be used.

Alternatively, the proportional-fair vector can be found using an iterative scheduling algorithm that maximizes the weighted sum rate during each frame and then updates the quality of service (QoS) weights appropriately. This algorithm is useful in practice for scheduling resources in time-varying channels (Section 5.5.4). While we discuss this algorithm in the context of time-invariant channels, a more general gradient scheduling algorithm is asymptotically optimal in time-varying channels [149].

For a given set of channel realizations, the algorithm operates iteratively over multiple frames. During frame f, the scheduler allocates resources to achieve rate vector $\mathbf{R}^{(f)}$ where

$$\mathbf{R}^{(f)} = \arg\max_{\mathbf{R} \in \mathcal{C}} \sum_k q_k^{(f)} R_k \tag{5.27}$$

and the QoS weight $q_k^{(f)}$ is a non-negative scalar. If we let the QoS weight for the kth user be the reciprocal of the smoothed average rate:

$$q_k^{(f)} = \frac{1}{\bar{R}_k^{(f)}}, \tag{5.28}$$

where

$$\bar{R}_k^{(f)} = \alpha \bar{R}_k^{(f-1)} + (1 - \alpha) R_k^{(f-1)}, \tag{5.29}$$

and $\alpha \in [0, 1]$ is a forgetting factor, then the rate vector $\mathbf{R}^{(f)}$ converges to the proportional-fair vector for sufficiently small α and sufficiently large f [149]. We note that the rate vector that maximizes the weighted sum rate in (5.27) can, in general, service multiple users if multiuser MIMO techniques are used.

If the transmission is restricted to serve a single user, then the solution to the weighted sum rate problem is to serve the user with the largest weighted rate. For single-user transmission, the achievable rate during frame f is

$$R^{(f)} = \arg\max_{R \in \mathcal{C}} \frac{R_k^{(f)}}{\bar{R}_k^{(f)}}. \tag{5.30}$$

The generalization of this procedure to time-varying channels is known as the proportional-fair scheduling algorithm for single-user service [150].

In delay-intolerant applications such as streaming media, a minimum average rate could be required to meet QoS targets. The proportional-fair criterion in (5.24) could be modified to ensure a minimum rate $R^{\min} > 0$ is achieved by all users:

$$\mathbf{R}^{\mathrm{Opt}} = \arg\max_{\mathbf{R} \in \mathcal{C}, R_k \geq R^{\min} \forall k} \sum_k \log R_k. \tag{5.31}$$

The optimal $\mathbf{R}^{\mathrm{Opt}}$ satisfying (5.31) can be found using the same iterative weighted sum-rate maximization in (5.27) but using a modified QoS weight that depends on a token counter $T_k^{(f)}$ [151]:

$$q_k^{(f)} = \frac{\exp\left(a_k T_k^{(f)}\right)}{\bar{R}_k^{(f)}}, \tag{5.32}$$

where $a_k > 0$ is an aggressiveness parameter. This token counter is updated each frame by incrementing it with the minimum desired rate $R^{\min}$ and re-

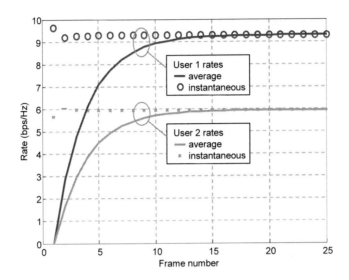

Fig. 5.19 Instantaneous and average rates for a $K = 2$-user broadcast channel under proportional-fair scheduling using DPC. The capacity region is the right subfigure of Figure 5.18. The initial rate vector achieved during frame 1 corresponds to the sum-rate capacity. The steady-state rate vector is the proportional-fair rate vector which maximizes the sum of log rates.

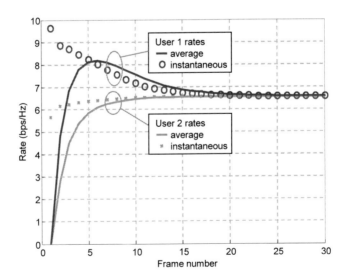

Fig. 5.20 Instantaneous and average rates for a $K = 2$-user broadcast channel under proportional-fair scheduling with minimum rate constraints. The constraints are set so they are outside the achievable rate region. As a result, the rates converge to the equal-rate vector.

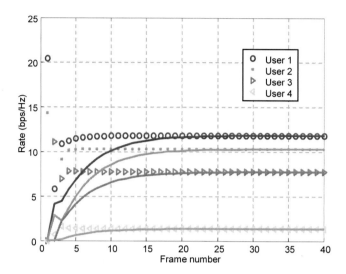

Fig. 5.21 Instantaneous and average rates for a $(4,4(1))$ time-invariant broadcast channel under proportional-fair scheduling and DPC. Instantaneous rates reach their steady-state values after a few frames.

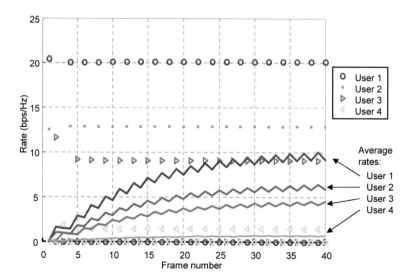

Fig. 5.22 Instantaneous and average rates using ZF for the same $(4,4(1))$ channel used in Figure 5.21. During steady-state operation, the transmitter serves users 1 and 2 during even-numbered frames and users 3 and 4 during odd-numbered frames. Only half of the $M = 4$ spatial degrees of freedom are used during each frame.

ducing it by the served rate:

$$T_k^{(f)} = \max\left(0, T_k^{(f-1)} + R_{min} - R_k^{(f-1)}\right). \tag{5.33}$$

Using (5.32) in (5.27), the iterative algorithm converges to the optimal $\mathbf{R}^{\text{Opt}}$ satisfying (5.31) if that rate vector is achievable [149].

Figure 5.19 shows the instantaneous and average rates for 2-user broadcast channel using DPC and proportional-fair scheduling. The capacity region of this channel is shown in Figure 5.18. During the first frame, the QoS weights for the two users are initialized to the same value. Since $q_1 = q_2$, then from (5.27), the rate vector achieved during this frame maximizes the sum rate. On subsequent frames, the QoS weights are updated according to (5.28) and (5.29), and the average rates converge to the proportional-fair rates.

Figure 5.20 shows the instantaneous and average rates for the same 2-user channel using DPC and proportional-fair scheduling with minimum rate requirements. The minimum rate R_{min} for both users is set to be outside the achievable rate region. Because this rate vector is not achievable, the average rates converge to the equal-rate vector.

Figure 5.21 shows the instantaneous and average rates for a time-invariant broadcast channel with $M = 4$ antennas and $K = 4$ single-antenna users using DPC. The sum-rate maximizing rate vector is achieved during frame 1 because the QoS weights are the same for all users. Even though there are four spatial degrees of freedom, the sum rate is maximized by serving only two users. When the performance reaches its steady state proportional-fair rate vector, all users are served each frame. Figure 5.22 shows the scheduled rates for the same channel using zero-forcing (Section 4.2.1). The users' channels are linearly independent so that up to four users could be served at once. However, in steady state, only two users are served during each frame, and the proportional-fair rate vector is achieved by time-multiplexing service between two pairs of users. While user 1 achieves 20 bps/Hz whenever it is served, it achieves this rate only half the time, so its average rate of 10 bps/Hz is below that of its proportional-fair rate achieved under DPC (about 12 bps/Hz).

5.4.3 Methodology for proportional-fair criterion

Under the proportional-fair criterion for the downlink, bases transmit with full power and user rates are adjusted so that the proportionally-fair rate vector is achieved by each base. The same principle applies to the uplink, except that users are power controlled to prevent excessive interference.

Figure 5.23 shows an overview of the proportional-fair methodology. For a given iteration n, the location of users, shadowing realizations, and channel fading realizations are fixed. An inner loop, running the iterative scheduling algorithm described in Section 5.4.2 over many frames, provides the proportional-fair rate vector $(R_{1,b}^{(n)}, \ldots, R_{K,b}^{(n)})$ for iteration n and base $b = 1, \ldots, B$. This methodology can be generalized for the case of time varying channels by allowing the channel fading to evolve from frame to frame.

5.4.3.1 Metrics

The performance metrics for the proportional-fair criterion are the mean throughput per site, the peak user rate, and the cell-edge user rate. The throughput measures the average data rate transmitted or received through the S sectors of a cell site. The peak and cell-edge rates measure the performance from the user's perspective.

Proportional-fair rate vectors are collected across the bases and over multiple iterations to generate the performance metrics. Each base serves K users, and each base is associated with a sector. Therefore with B bases in the network and S sectors per site, there are B/S sites in the network. Given the user rates $R_{1,b}^{(n)}, \ldots, R_{K,b}^{(n)}$ for bases $b = 1, \ldots, B$ and iterations $n = 1, \ldots, Q$, we define the *mean throughput per site* as

$$\frac{1}{Q(B/S)} \sum_{b=1}^{B} \sum_{k=1}^{K} \sum_{n=1}^{Q} R_{k,b}^{(n)} = SK \frac{1}{KQB} \sum_{b=1}^{B} \sum_{k=1}^{K} \sum_{n=1}^{Q} R_{k,b}^{(n)}. \qquad (5.34)$$

The expression on the right shows that the mean throughput per site is the mean user rate multiplied by the number of total users in the network SK. By treating the user rates as realizations of a random variable R, we define the *peak user rate* as the rate z such that the probability of R being less than z is 90%. Using the inverse cumulative distribution function, the peak user rate is $F_R^{-1}(0.9)$. Similarly, the *cell-edge user rate* is the rate z such that the probability of R being less than z is 10%: $F_R^{-1}(0.1)$. Since the mean throughput per site is SK times the mean user rate, we can make easier comparisons with the cell-edge and peak user rates by normalizing these values. We define the *normalized peak user rate* as $SK \times F_R^{-1}(0.9)$ and the *normalized cell-edge rate* as $SK \times F_R^{-1}(0.1)$.

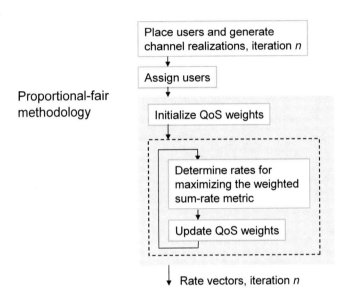

Fig. 5.23 Overview of the proportional-fair simulation methodology. For iteration n, the proportional-fair rate vector is found for the K users served by each base $b = 1, \ldots, B$. $(R_{1,b}, \ldots, R_{K,b}) \in \mathcal{C}_b$. The throughput and user rate metrics are computed from rate vectors collected over many iterations.

5.4.3.2 Downlink

For a given network of B bases, we place a user indexed by k randomly with a uniform spatial distribution in the network. With respect to each base $b = 1, \ldots, B$, we measure the distance $d_{k,b}$ to the user, determine the directional antenna gain $G_{k,b}$, and generate a random shadowing realization $Z_{k,b}$. All bases are assumed to transmit with full power P, and the user is assigned to the base with the highest SNR, or equivalently, the highest channel gain (5.5). The serving base, which we denote with index b^*, is given by

$$b^* = \arg\max_b \alpha_{k,b}^2 = \arg\max_b \left(\frac{d_{k,b}}{d_{\text{ref}}}\right)^{-\gamma} Z_{k,b} G_{k,b}, \tag{5.35}$$

where the additional maximization with respect to the wraparound replicas (Section 5.2.3) has not been explicitly stated. We repeat this process of placing and assigning users until each base is assigned K users. If a user is assigned to a base with K users already, then the user is simply discarded.

As explained in Section 1.2.2, the user *geometry* is an important metric for characterizing the link performance in systems with co-channel interference. The geometry is defined as the ratio between the average power received from the desired source and the average noise plus interference power. From (5.4), the downlink geometry for user k assigned to base b^* is

$$\frac{P\alpha^2_{k,b^*}}{\sigma^2 + \sum_{b \neq b^*} P\alpha^2_{k,b}} = \frac{(P/\sigma^2)\alpha^2_{k,b^*}}{1 + \sum_{b \neq b^*} (P/\sigma^2)\alpha^2_{k,b}}, \tag{5.36}$$

where the expression on the right gives the geometry in terms of the reference SNR P/σ^2. The geometry is a random variable that depends on the location and shadowing realization of user k with respect to all the bases. As we will see in Section 6.3, in a typical cellular network with universal reuse, the geometry distribution has a typical range of -10 dB to 20 dB. If we treat the interference as additive noise, then the relevant range of SNRs for the equivalent isolated SU- or MU-MIMO channel is likewise -10 dB to 20 dB.

For a given placement of users and channel realizations, the proportional-fair rate vectors for the B bases are determined using the iterative scheduling algorithm from Section 5.4.2. Each step of this algorithm requires determining the rate vector that maximizes the weighted sum rate given the current QoS weights. Given the received signal model (5.4), the transmit covariances $\mathbb{E}(\mathbf{s}_b\mathbf{s}_b^H)$ of the interfering bases $b \neq b^*$ are not known, and therefore it is not possible to determine the optimal transmit covariance for base b^* to maximize the weighted sum rate. Likewise, the transmit covariance for the other bases cannot be determined without knowledge of their respective interfering bases. This chicken-and-egg problem is addressed by first computing the transmit covariance of the serving base assuming the signal from the other bases $b \neq b^*$ is transmitted isotropically and averaged over the Rayleigh fading. The covariance of the interference received from base b is

$$\mathbb{E}\left(\mathbf{H}^H_{k,b}\mathbf{s}_b\mathbf{s}_b^H\mathbf{H}_{k,b}\right) = \frac{P}{M}\mathbb{E}\left(\mathbf{H}^H_{k,b}\mathbf{H}_{k,b}\right) = P\alpha^2_{k,b}\mathbf{I}_N. \tag{5.37}$$

Therefore we can rewrite the received signal (5.4) as

$$\mathbf{x}_k = \mathbf{H}^H_{k,b^*}\mathbf{s}_{b^*} + \tilde{\mathbf{n}}_k, \tag{5.38}$$

where $\tilde{\mathbf{n}}_k$ is the combined noise and interference modeled as a zero-mean Gaussian vector with covariance

$$\mathbb{E}\left(\tilde{\mathbf{n}}_k \tilde{\mathbf{n}}_k^H\right) = \sigma^2 \mathbf{I}_N + \mathbf{I}_N \sum_{b \neq b^*} \alpha_{k,b}^2 P. \tag{5.39}$$

The transmit covariance $\mathbf{Q}_{b^*}$ can be calculated as if base b^* were serving its K users in ZMSW noise with covariance (5.39). For example, for CL SU-MIMO transmission, the transmit covariance for base b^* is

$$\mathbf{Q}_{b^*} = \arg \max_{\mathbf{Q}, tr\mathbf{Q} \leq P} \log_2 \det \left[\mathbf{I}_N + \frac{1}{\sigma^2 + \sum_{b \neq b^*} P\alpha_{k,b}^2} \mathbf{H}_{k,b^*}^H \mathbf{Q} \mathbf{H}_{k,b^*}\right]. \tag{5.40}$$

The covariances of the other bases $\mathbf{Q}_b, b = 1, \ldots, B, b \neq b^*$ can be determined in a similar manner.

Given the covariance $\mathbf{Q}_b$ for base b, the interference received by user k from base b, as a function of the channel realization $\mathbf{H}_{k,b}$, has covariance $\mathbf{H}_{k,b} \mathbf{Q}_b \mathbf{H}_{k,b}^H$. The combined noise and interference in (5.4) has covariance

$$\mathbf{Q}_{I,k} := \sigma^2 \mathbf{I}_N + \sum_{b \neq b^*} \mathbf{H}_{k,b}^H \mathbf{Q}_b \mathbf{H}_{k,b}. \tag{5.41}$$

Whereas the covariance in (5.39) is averaged over the channel realizations, the covariance in (5.41) is a function of the block-fading channel realizations.

The interference covariance in (5.41) is assumed to be known at the receiver. Therefore the achievable rate for users served by base b^* can be computed based on the whitened received signal $\mathbf{Q}_{I,k}^{-1/2} \mathbf{x}_k$, given the transmit covariance $\mathbf{Q}_{b^*}$ determined in the presence of averaged interference. For the example of SU transmission with covariance $\mathbf{Q}_{b^*}$ computed from (5.40), the served mobile achieves rate

$$R = \log_2 \det \left[\mathbf{I}_N + \mathbf{Q}_{I,k}^{-1} \mathbf{H}_{k,b^*}^H \mathbf{Q}_{b^*} \mathbf{H}_{k,b^*}\right], \tag{5.42}$$

where $\mathbf{Q}_{I,k}$ is from (5.41). The average rate (5.42), taken with respect to realizations of the interferers' channels ($\mathbf{H}_{k,b}, b \neq b^*$), can be lowered bounded as follows:

$$\mathbb{E}_{\mathbf{H}_{k,b}}(R) \geq \log_2 \det \left[\mathbf{I}_N + \frac{1}{\sigma^2 + \sum_{b \neq b^*} P\alpha_{k,b}^2} \mathbf{H}_{b^*}^H \mathbf{Q}_{b^*} \mathbf{H}_{b^*}\right]. \tag{5.43}$$

This bound can be obtained by extending the derivation for (5.16) and applying Jensen's inequality to the convex function

$$f(\mathbf{A}) = \log_2 \det \left(\mathbf{I} + (\mathbf{I} + \mathbf{A}^H \mathbf{R} \mathbf{A})^{-1} \mathbf{X}\right). \tag{5.44}$$

This inequality shows that the rate averaged with respect to the interference with covariance from (5.41) will be no less than the rate achieved under the average interference with covariance from (5.39).

In summary, the downlink transmit covariance for the proportional-fair criterion is set under the assumption of averaged interference (5.39), but the rate is computed using this transmit covariance and interference with covariance (5.41).

5.4.3.3 Uplink

On the downlink, users are assigned to the base with the lowest pathloss. As a consequence, because bases transmit with the same power, the signal received by a user from its assigned base is always stronger than that from any interfering base. On the uplink, users are also assigned to the base with the lowest pathloss. However, as a result of shadowing, an interfering user could be received at a base more strongly than its assigned user, resulting in a very low geometry. Alternatively, the SINR could be very high if the desired user is very close to its assigned base. In the interest of fairness, large variations in geometry are not desirable, and *power control* is used to adjust the transmit powers so that the geometry is near a nominal target ρ.

Due to power control, the transmit power for user k assigned to base b ($k \in U_b$) is $\gamma_k P_k$, where $\gamma_k \in [0, 1]$ is the fraction of the power constraint P_k. Fixing the power fraction of users assigned to other bases $\gamma_j, j \notin U_b$, the geometry of user k is

$$\Gamma_k(\gamma_k) = \frac{\gamma_k P_k \alpha_{k,b}^2}{\sigma^2 + \sum_{j \notin U_b} \gamma_j P_j \alpha_{j,b}^2}, \qquad (5.45)$$

where $\alpha_{j,b}^2$ is the channel gain (5.5) between user j and base b. The power control algorithm adjusts the powers so that users with favorable channels achieve the target geometry ρ transmitting power $\gamma_k P_k$ (with $\gamma_k < 1$), while users with unfavorable channels achieve a geometry less than ρ even though they transmit with full power P_k. For user k, the power fraction is set by fixing the power fractions for other users and determining the required fraction γ_k to achieve ρ or setting $\gamma_k = 1$ if ρ is not achievable:

$$\gamma_k = \begin{cases} \frac{\rho\left(\sigma^2 + \sum_{j \notin U_b} \gamma_j P_j \alpha_{j,b}^2\right)}{P_k \alpha_{k,b}^2} & \text{if } \Gamma_k(1) \geq P_k \\ 1 & \text{if } \Gamma_k(1) < P_k. \end{cases} \qquad (5.46)$$

Power fractions are adjusted iteratively over all users until the values converge.

If a user's gain to its assigned base is much greater than the gain to the other bases, it could potentially transmit with more power and increase its geometry without causing significant interference. To account for this possibility, we propose a more aggressive power control scheme which sets the new target geometry to be

$$\rho \left(\frac{\alpha_{k,b}^2}{\alpha_{k,\bar{b}}^2} \right)^\beta , \tag{5.47}$$

where ρ is the nominal target, and $\alpha_{k,\bar{b}}^2$ is the channel gain between user k and base $\bar{b}$ which has the *next* highest channel gain after its assigned base b. The parameter $\beta \geq 0$ adjusts the aggressiveness of the new target. Setting $\beta = 0$ does not change the target geometry. A higher value of β gives a more aggressive boost, resulting in a higher peak geometry but perhaps at the expense of a lower tail.

Figure 5.24 shows the CDF of the uplink geometry (5.45) for the case of universal reuse, $S = 3$ sectors per cell site, $K = 1$ user per sector, and reference SNR $P/\sigma^2 = 30$ dB. Using power control with a nominal target SINR of $\rho = 6$ dB and an aggressiveness parameter $\beta = 0.5$, the range of the geometry is significantly reduced compared to the case with no power control where all users transmit with full power. In particular, the minimum geometry with power control is about -14 dB, while it is less than -20 dB without power control.

Once the power fractions γ_k are set by the power control algorithm offline, the uplink simulation methodology follows the downlink methodology described above in Section 5.4.3.2, except of course that the multiple access channel model is used instead of the broadcast channel.

The proportional-fair methodology, corresponding to the gray box in Figure 5.23, can be summarized as follows for both the downlink and uplink:

Proportional-fair algorithm

1. **Initialize**
   Set $f := 0$. Set QoS weights $q_k^{(f)} = 1/\epsilon$, $(\epsilon \ll 1)$ for all $k = 1, \ldots, KB$.
2. For $f \geq 0$, do the following:
   a. **Update transmit covariances**
      Assuming average interference, determine transmit covariances to

maximize the weighted sum rate.

For the downlink, update base covariances $\mathbf{Q}_b^{(f)}$, $b = 1, \ldots, B$.

For the uplink, update user covariances $\mathbf{Q}_k^{(f)}$, $k = 1, \ldots, KB$.

b. **Determine achievable rate vector**

In the presence of interference based on the covariances computed in the previous step and using the instantaneous channel realizations, determine the rate vector for each base: $\mathbf{R}_{1,b}^{(f)}, \ldots, \mathbf{R}_{K,b}^{(f)}$, for $b = 1, \ldots, B$.

c. **Update QoS weights**

Update $q_k^{(f+1)}$ for $k = 1, \ldots, KB$, using, for example, (5.28).

d. Let $f := f + 1$ and iterate until convergence.

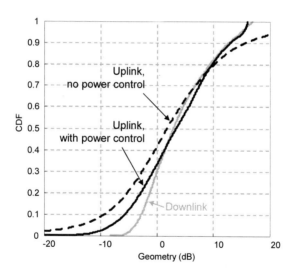

Fig. 5.24 CDF of uplink geometry (5.45) with and without power control. Power control reduces the range of the upper and lower tails, making the distribution similar to the downlink geometry distribution (Figure 6.15).

5.4.4 Methodology for equal-rate criterion

Under the equal-rate methodology, the performance is determined by the achievable rate of the user with the poorest channel conditions. To prevent the performance from being dominated by a few users with very poor channel conditions, 10% of the users with the worst channels are declared to be in outage and receive no service. Transmit powers are adjusted so that all users in the network, except those in outage, receive a common data rate. Note that under the proportional-fair methodology, users are not explicitly placed in outage, and those with poor channels receive lower rates as a result of the proportional-fair criterion.

The goal is to determine the largest common data rate that is consistent with the target user outage probability of 10%. Figure 5.25 shows an overview of the equal-rate methodology. For a given target rate R^* and for a given iteration f where the location of users and channel realizations are fixed, an inner loop calculates the fraction of users $u_{R^*}^{(f)}$ in outage due to unfavorable channel and/or interference conditions. The outage probability for R^* is determined by averaging $u_{R^*}^{(f)}$ over many iterations. The desired equal-rate performance metric is the rate R^* corresponding to the average outage probability of 10%.

While the equal-rate rate vector can be found using the iterative scheduling algorithm (Section 5.4.2) with no users in outage, we describe a different technique below that accounts for user outage. The technique determines the outage probability for a few target rates R^* and interpolates between these rates to obtain the rate that results in 10% user outage. Below we describe the details for the uplink and downlink channels where, for simplicity, we consider linear transmitter and receiver processing and single-antenna terminals.

5.4.4.1 Uplink

We first present the uplink algorithm for the conventional network where bases operate independently. We then present an extension for network MIMO with coordinated bases.

Let the target rate for each user in the network be R^* bps/Hz. Assuming Gaussian signaling and ideal coding, the target rate of R^* bps/Hz translates to a target SINR of $\rho \triangleq 2^{R^*} - 1$ for each user. Suppose that the target SINR ρ is small enough for all the users to achieve it, given the power constraint on each user and the interference between users. This means that there exists a

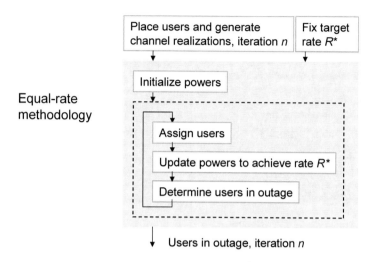

Fig. 5.25 Overview of equal-rate simulation methodology. For iteration n and for a given target rate R^*, the outage probability is computed across users in the entire network. Given R^*, the average outage probability is computed over many iterations. The target rate corresponding to an average outage probability of 10% is determined by varying R^*.

feasible setting of each user's transmitted power, and an assignment of users to bases, such that each user attains an SINR of ρ or higher at its assigned base, with an SINR-maximizing linear MMSE receiver.

If we denote the base assigned to user k as b_k and the transmit power of user k as p_k, the SINR of user k is

$$p_k \mathbf{h}_{k,b_k}^H \left(\sigma^2 \mathbf{I}_M + \sum_{j \neq k} p_j \mathbf{h}_{j,b_k} \mathbf{h}_{j,b_k}^H \right)^{-1} \mathbf{h}_{k,b_k}. \tag{5.48}$$

Note that this expression assumes perfect knowledge at base b_k of the channel vector $\mathbf{h}_{k,b_k}$ and the composite interference covariance $\sum_{j \neq k} p_j \mathbf{h}_{j,b_k} \mathbf{h}_{j,b_k}^H$. We define an optimization problem for minimizing the total power as follows:

$$\min_{\{b_k\},\{p_k\}} \sum_{k=1}^{KB} p_k, \tag{5.49}$$

subject to

$$p_k \mathbf{h}_{k,b_k}^H \left(\sigma^2 \mathbf{I}_M + \sum_{j \neq k} p_j \mathbf{h}_{j,b_k} \mathbf{h}_{j,b_k}^H \right)^{-1} \mathbf{h}_{k,b_k} \geq \rho, \quad p_k \geq 0. \qquad (5.50)$$

For this problem, the following iterative algorithm from [126] (also see [152, 153]) can be used to determine the transmitted powers and base assignments for all the users. Letting $p_k^{(f)}$ represent the uplink power of user k in the fth frame, we first initialize all powers to zero: $p_k^{(0)} = 0$ for all k. For the fth frame, given the set of powers $\{p_k^{(f)}\}$, we assign each user k to its best base and update its transmit power. The user is assigned to the base where it would attain the highest SINR:

$$b_k = \arg \max_{b=1,\ldots,B} \mathbf{h}_{k,b}^H \left(\sigma^2 \mathbf{I}_M + \sum_{j \neq k} p_j^{(f)} \mathbf{h}_{j,b} \mathbf{h}_{j,b}^H \right)^{-1} \mathbf{h}_{k,b}. \qquad (5.51)$$

The power for user k is updated for frame $f + 1$ so it achieves the target SINR ρ at the assigned base b_k, assuming every other user j transmits at the current power level of $p_j^{(f)}$:

$$p_k^{(f+1)} = \rho \left[\mathbf{h}_{k,b_k}^H \left(\sigma^2 \mathbf{I}_M + \sum_{j \neq k} p_j^{(f)} \mathbf{h}_{j,b} \mathbf{h}_{j,b}^H \right)^{-1} \mathbf{h}_{k,b_k} \right]^{-1}. \qquad (5.52)$$

The assignment and power updates are repeated for all other users, and the entire procedure is repeated until convergence is achieved.

The algorithm above has been shown to converge [126] to transmitted powers $\{\tilde{p}_k\}$ that not only minimize the sum powers in (5.49), but are optimal in an even stronger sense: if it is possible for every user to attain the target SINR of ρ with transmitted powers $\{p_k\}$, then $p_k \geq \tilde{p}_k$ for every k. In other words, the algorithm minimizes the power transmitted by *every* user, subject to the target SINR of ρ being achieved by all users.

In general, it might be impossible for all the users to achieve the target SINR simultaneously. It is then necessary to settle for serving only a subset of the users, declaring the rest to be in outage. (This might be preferable to serving all the users at a low rate determined by the user with the worst channel conditions.) In principle, the largest supportable subset of users could be determined by sequentially examining all subsets of users in decreasing order of size, but this approach is practical only when the number of users is small.

Instead, we will modify the iterative algorithm of [126] slightly to obtain a suboptimal but computationally efficient algorithm for determining which subset of users should be served. After each frame f, the modified algorithm declares users whose updated powers exceed the power constraint of 1 to be in outage, and eliminates them from consideration in future frames. This progressive elimination of users eventually results in a subset of users that can all simultaneously achieve the target SINR ρ. For this subset of users, the algorithm then finds the optimal transmitted powers and cluster assignments. However, the user subset itself need not be the largest possible; essentially, this is because we do not allow a user consigned to outage to be resurrected in a future frame.

The algorithm described above applies to a conventional cellular network where bases operate independently. It can be extended to the case of uplink network MIMO where clusters of bases work jointly to demodulate a user's signal. A *coordination cluster* is defined to be a subset of base stations that jointly serve one or more users through the antennas of all their sectors. The network is postulated to have a predefined set of coordination clusters, and each user can be assigned to any one of these clusters.

To highlight the dependence of the spectral efficiency gain on the number of coordinated bases, coordination clusters with a specific structure are of interest. For any integer $r \geq 0$, an *r-ring coordination cluster* is defined to consist of any base station and the first r rings of its neighboring base stations (accounting for wraparound), and $\mathcal{C}_r$ is defined to be the set of all r-ring coordination clusters in the network. Figure 5.26 illustrates r-ring clusters for $r = 0, 1, 2, 4$.

With $\mathcal{C}_0$ as the set of coordination clusters in the network, there is in fact no coordination between base stations. This case serves as the benchmark in estimating the spectral efficiency gain achievable with sets of larger coordination clusters.

With some abuse of notation, let $\mathbf{h}_{k,C} \in \mathbb{C}^{3N|C|}$ denote the channel from user k to the antennas of all the base stations in the coordination cluster C; here $|C|$ denotes the number of base stations in C, and the factor of 3 comes from assuming $S = 3$ sectors per site. Then, with user k transmitting power p_k, the SINR attained by user k at cluster C is $\mathbf{h}_{k,C}^* \left(\sigma^2 \mathbf{I}_M + \sum_{j \neq k} p_j \mathbf{h}_{j,C} \mathbf{h}_{j,C}^* \right)^{-1} \mathbf{h}_{k,C} \, p_k$. Using this expression, we can generalize the original algorithm for conventional networks to the case of network MIMO in a straightforward manner by replacing the user assignment

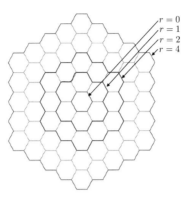

Fig. 5.26 Coordination clusters with r rings. The case of $r = 0$ denotes intrasite coordination where the sector antennas at a site are coordinated.

(5.51) with

$$
C_k^{(f)} = \arg \max_{C \in \mathcal{C}_r} \mathbf{h}_{k,C}^H \left(\sigma^2 \mathbf{I}_M + \sum_{j \neq k} p_j^{(f)} \mathbf{h}_{j,C} \mathbf{h}_{j,C}^H \right)^{-1} \mathbf{h}_{k,C}, \qquad (5.53)
$$

and by replacing the power update (5.52) with

$$
p_k^{(f+1)} = \rho \left[\mathbf{h}_{k,C_k^{(f)}}^H \left(\sigma^2 \mathbf{I}_M + \sum_{j \neq k} p_j^{(f)} \mathbf{h}_{j,C_k^{(f)}} \mathbf{h}_{j,C_k^{(f)}}^H \right)^{-1} \mathbf{h}_{k,C_k^{(f)}} \right]^{-1}. \qquad (5.54)
$$

The uplink equal-rate methodology, corresponding to the gray box in Figure 5.25, is summarized below.

Uplink equal-rate algorithm

1. **Initialize**

   Set $f := 0$, start with no users in outage, and let $p_k^{(f)} = 0$ for all $k = 1, \ldots, KB$.

2. For frame $f \geq 0$, given $\left\{ p_k^{(f)} \right\}$, do the following:

   a. **Assign user**

      Assign each user $(k = 1, \ldots, KB)$ to the base/cluster where it would attain the highest SINR: (5.51) for a conventional network or (5.53) for network MIMO.

b. **Update transmit power**

Update for each user $(k = 1, \ldots, KB)$ the uplink power $p_k^{(f+1)}$:
(5.52) for conventional or (5.54) for network MIMO.

c. **Determine users in outage**

For each base $(b = 1, \ldots, B)$ and for each user $k \in U_b$, if
$p_k^{(f+1)} > 1$, eliminate the user and reset its power to zero in all
future frames.

d. Let $f := f + 1$ and iterate until convergence.

5.4.4.2 Downlink

As in the uplink, all users in the downlink equal-rate methodology, except
those in outage, must be served at a common data rate R^* with corresponding SINR ρ. As before, our goal is to determine the largest common rate that
is consistent with the desired user outage probability. However, unlike the uplink where the interference covariance and CSI of all users is known at a given
receiving base—this assumption is reasonable since the covariances could be
estimated from the interfering users' pilots signals—the downlink CSI of all
users is not necessarily known at any given base. We consider the following
three transceiver architectures: single-base precoding with local CSIT, coordinated precoding with global CSIT, and network MIMO precoding with
global CSIT and global data.

Under the assumption of linear precoding, the transmitted signal from
base b can be written as

$$\mathbf{s}_b = \sum_{j \in U_b} \sqrt{v_{j,b}} \mathbf{g}_{j,b} u_j, \tag{5.55}$$

where U_b is the set of users served by base b, $v_{j,b}$ is the transmit power
allocated by base b to user j, $\mathbf{g}_{j,b} \in \mathbb{C}^M$ is the corresponding normalized
beamforming vector with unit norm, and u_j is the unit-power information-
bearing signal for user j. We will denote by b_k the base station to which
user k is assigned. Using (5.55) in (5.4), the received signal for user k under
conventional network operation can then be written as

$$x_k = \mathbf{h}_{k,b_k}^H \mathbf{s}_{b_k} + \sum_{b \neq b_k} \mathbf{h}_{k,b}^H \mathbf{s}_b + n_k \tag{5.56}$$

$$= \sqrt{v_{k,b_k}} \mathbf{h}_{k,b_k}^H \mathbf{g}_{k,b_k} u_k + \sum_{j \in U_{b_k}, j \neq k} \sqrt{v_{j,b_k}} \mathbf{h}_{k,b_k}^H \mathbf{g}_{j,b_k} u_j + \quad (5.57)$$

$$\sum_{b \neq b_k} \sum_{j \in U_b} \sqrt{v_{j,b}} \mathbf{h}_{k,b}^H \mathbf{g}_{j,b} u_j + n_k. \quad (5.58)$$

Given the beamforming vectors for all users, the SINR of user k assigned to base b_k is

$$\frac{|\mathbf{h}_{k,b_k}^H \mathbf{g}_{k,b_k}|^2 v_{k,b_k}}{\sigma^2 + \sum_{j \in U_{b_k}, j \neq k} |\mathbf{h}_{k,b_k}^H \mathbf{g}_{j,b_k}|^2 v_{j,b_k} + \mu_k}, \quad (5.59)$$

where

$$\mu_k = \sum_{b \neq b_k} \sum_{j \in U_b} v_{j,b} |\mathbf{h}_{k,b}^H \mathbf{g}_{j,b}|^2 \quad (5.60)$$

is the intercell interference received from the other bases. The downlink optimization problem for the conventional network can be stated as follows:

$$\min_{\{b_k\},\{\mathbf{g}_k\},\{v_k\}} \sum_{k=1}^{KB} v_k \quad (5.61)$$

subject to

$$\frac{|\mathbf{h}_{k,b_k}^H \mathbf{g}_{k,b_k}|^2 v_{k,b_k}}{\sigma^2 + \sum_{j \neq k, j \in U_{b_k}} |\mathbf{h}_{k,b_k}^H \mathbf{g}_{j,b_k}|^2 v_{j,b_k} + \mu_k} \geq \rho, \quad (5.62)$$

$$\|\mathbf{g}_{k,b_k}\| = 1, \quad (5.63)$$

$$v_{k,b_k} \geq 0, \quad (5.64)$$

for $k = 1, \ldots, KB$, where μ_k is given by (5.60). In contrast to the uplink optimization where the SINR (5.48) could be expressed implicitly as a function of the MMSE receiver, the downlink SINR (5.59) requires the beamforming vector to be given explicitly.

To solve this optimization we use the linear BC/MAC duality from Section 4.3: if a given rate vector is achievable in one direction for given sets of base assignments and beamforming weights, then it is achievable in the other direction with the same base assignments, same beamforming weights, and same sum power. We could therefore run the dual uplink algorithm until convergence. Then, fixing the base assignments and beamforming weights, iterate until downlink powers converge. In theory this procedure would require dual infinite frame iterations. A more elegant algorithm based on [124] would run the frames in parallel so that the downlink powers $\{q_k^{(f)}\}$ during frame f are updated immediately after the fth frame of the algorithm for determining the dual uplink powers $\{p_k^{(f)}\}$. The parallel algorithm can be shown to converge to the solution that minimizes the sum of all transmitted powers. In contrast to the uplink, however, it is in general not possible to minimize the individual user powers. Normalizing the channel so the total interference plus noise power is unity, the downlink received signal (5.56) for user $k = 1, \ldots, KB$ can be written as

$$x_k = \frac{1}{\sqrt{\sigma^2 + \mu_k}} \mathbf{h}_{k,b}^H \mathbf{s}_b + n'_k, \tag{5.65}$$

where n'_k has unit variance. The dual uplink received signal at base $b = 1, \ldots, B$ is therefore:

$$\mathbf{x}_b = \sum_{k \in U_b} \frac{1}{\sqrt{\sigma^2 + \mu_k}} \mathbf{h}_{k,b} s_k + \mathbf{n}'_b, \tag{5.66}$$

where n'_b has unit variance. The downlink algorithm iteratively updates the user assignment, the MMSE beamforming vector for the dual uplink, the downlink interference power, the dual uplink transmit powers, and the downlink transmit powers.

The downlink algorithm is initialized with no users in outage and with $q_k^{(0)} = p_k^{(0)} = 0$ for all k. The interference is initialized to $\mu^{(-1)} = 0$. During frame f, user k is assigned to the base where it would attain the highest dual uplink SINR (5.59):

$$b_k^{(f)} = \arg \max_{b=1,\ldots,B} \mathbf{h}_{k,b}^H \left(\mathbf{I}_M + \sum_{j \in U_b, j \neq k} \frac{p_j^{(f)}}{\sigma^2 + \mu_j^{(f-1)}} \mathbf{h}_{j,b} \mathbf{h}_{j,b}^H \right)^{-1} \mathbf{h}_{k,b}. \tag{5.67}$$

For each user $k \in U_b$, the corresponding unit-norm beamforming weight vector $\mathbf{g}_k^{(f)}$ derived from the MMSE criterion for the dual uplink is:

$$\mathbf{g}_k^{(f)} = \frac{\left(\mathbf{I}_N + \sum_{j \in U_b, j \neq k} \frac{p_j^{(f)}}{\sigma^2 + \mu_j^{(f-1)}} \mathbf{h}_{j,b} \mathbf{h}_{j,b}^H \right)^{-1} \mathbf{h}_{k,b}}{\left\| \left(\mathbf{I}_N + \sum_{j \in U_b, j \neq k} \frac{p_j^{(f)}}{\sigma^2 + \mu_j^{(f-1)}} \mathbf{h}_{j,b} \mathbf{h}_{j,b}^H \right)^{-1} \mathbf{h}_{k,b} \right\|}. \tag{5.68}$$

To simplify the notation, we have dropped the user and frame indices for the assigned base $b_k^{(f)}$. For each base $b = 1, \ldots, B$, and for each user $k \in U_b$, the interference power received from other bases $b' \neq b$ is updated:

$$\mu_k^{(f)} = \sum_{b' \neq b} \sum_{j \in U_{b'}} q_j^{(f)} \left| \mathbf{h}_{k,b'}^H \mathbf{g}_j^{(f)} \right|^2 . \tag{5.69}$$

Using the dual uplink received signal (5.66), the uplink power for user $k \in U_b$ is updated so that it achieves the target SINR ρ:

$$p_k^{(f+1)} = \frac{\rho \left(1 + \sum_{\substack{j \neq k \\ j \in U_b}} \frac{p_j^{(f)}}{\sigma^2 + \mu_j^{(f-1)}} \left| \mathbf{h}_{j,b}^H \mathbf{g}_k^{(f)} \right|^2 \right)}{\frac{\left| \mathbf{h}_{k,b}^H \mathbf{g}_k^{(f)} \right|^2}{\sigma^2 + \mu_k^{(f-1)}}} \tag{5.70}$$

Using the downlink SINR expression in (5.59), the downlink transmit power is updated for each user so that it achieves the target SINR ρ. For each base $b = 1, \ldots, B$, and for each $k \in U_b$, the power is set according to:

$$v_{k,b}^{(f+1)} = \frac{\rho \left(1 + \sum_{j \in U_b, j \neq k} |\mathbf{h}_{k,b}^H \mathbf{g}_j|^2 v_{j,b}^{(f)} + \mu_k \right)}{|\mathbf{h}_{k,b}^H \mathbf{g}_k^{(f)}|^2} . \tag{5.71}$$

The algorithm is repeated over many frames until convergence occurs. As was done for the uplink in the case of outage, we can modify the downlink algorithm to obtain a suboptimal but computationally efficient algorithm for determining the subset of served users. The basic idea is to check, after each frame, whether the power constraint at any sector antenna is violated and, if so, to eliminate users in decreasing order of requested power from that antenna until all power constraints are met. This approach is suboptimal because a user cannot be resurrected once assigned to the outage state. However, we use this technique because the optimal approach is prohibitively complex for typical network sizes.

We now discuss the related methodology for the case for interference nulling and network MIMO where CSI is known at all base stations. We describe the methodology for the more general case of network MIMO because interference nulling is a special case when each user's coordination cluster contains a single base. In general, we let C_k denote the set of bases belonging to the coordination cluster that serves user k. Note that given C_k for all $k = 1, \ldots, KB$, we can determine U_b, the set of users served by base b, for all $b = 1, .., B$, and vice versa. Under network MIMO, any base belonging to cluster C_k has knowledge of that user's data.

Given the coordination clusters C_k for all users $k = 1, \ldots, KB$, the received signal for a user k (5.3) can be written as

$$x_k = \sqrt{v_k}\mathbf{h}_{k,C_k}\tilde{\mathbf{g}}_k u_k + \sum_{j \neq k} \sqrt{v_j}\mathbf{h}_{k,C_j}\tilde{\mathbf{g}}_j u_j + n_k, \tag{5.72}$$

where $\mathbf{h}_{k,C_j} \in \mathbb{C}^{M|C_j|}$ is the vector of stacked channels from bases serving user j to the user k and where $\tilde{\mathbf{g}}_j \in \mathbb{C}^{M|C_j|}$ is the unit-norm vector of stacked beamforming vectors for the bases in the set C_j.

Given the beamforming vectors for all users in (5.72), the SINR of user k is

$$\frac{\left|\mathbf{h}_{k,C_k}^H \tilde{\mathbf{g}}_k\right|^2 v_k}{\sigma^2 + \sum_{j \neq k}\left|\mathbf{h}_{k,C_j}^H \tilde{\mathbf{g}}_j\right|^2 v_j}. \tag{5.73}$$

The downlink optimization problem under network MIMO can be stated as follows:

$$\min_{\{C_k\},\{\tilde{\mathbf{g}}_k\},\{v_k\}} \sum_{k=1}^{KB} v_k \tag{5.74}$$

subject to

$$\frac{\left|\mathbf{h}_{k,C_k}^H \tilde{\mathbf{g}}_k\right|^2 v_k}{\sigma^2 + \sum_{j \neq k}\left|\mathbf{h}_{k,C_j}^H \tilde{\mathbf{g}}_j\right|^2 v_j} \geq \rho, \quad \|\tilde{\mathbf{g}}_k\| = 1, \quad v_k \geq 0. \tag{5.75}$$

The iterative algorithm is similar to the previous downlink algorithm for the conventional network except that the channels do not need to be normalized by the noise plus interference power.

During frame f, user k is assigned to the base where it would attain the highest dual uplink SINR:

$$C_k^{(f)} = \arg\max_{C \in \mathcal{C}_r} \mathbf{h}_{k,C}^H \left(\sigma^2 \mathbf{I}_M + \sum_{j \neq k} p_j^{(f)} \mathbf{h}_{j,C} \mathbf{h}_{j,C}^H\right)^{-1} \mathbf{h}_{k,C}, \tag{5.76}$$

where the summation is taken over users j not in outage. The corresponding unit-norm beamforming weights $\tilde{\mathbf{g}}_k^{(f)}$ derived from the MMSE criterion for the dual uplink is

$$\tilde{\mathbf{g}}_k^{(f)} = \frac{\left(\sigma^2 \mathbf{I}_N + \sum_{j \neq k} p_j^{(f)} \mathbf{h}_{j,C_k^{(f)}} \mathbf{h}_{j,C_k^{(f)}}^H\right)^{-1} \mathbf{h}_{k,C_k^{(f)}}}{\left\|\left(\sigma^2 \mathbf{I}_N + \sum_{j \neq k} p_j^{(f)} \mathbf{h}_{j,C_k^{(f)}} \mathbf{h}_{j,C_k^{(f)}}^*\right)^{-1} \mathbf{h}_{k,C_k^{(f)}}\right\|}. \tag{5.77}$$

Update the dual uplink power $p_k^{(f+1)}$ and downlink power $q_k^{(f+1)}$ for user k to attain the target SINR ρ assuming all other powers are fixed:

$$p_k^{(f+1)} = \frac{\rho \left[\sigma^2 + \sum_{j \neq k} p_j^{(f)} \left| \mathbf{h}_{j,C_k^{(f)}}^H \tilde{\mathbf{g}}_k^{(f)} \right|^2 \right]}{\left| \mathbf{h}_{k,C_k^{(f)}}^H \tilde{\mathbf{g}}_k^{(f)} \right|^2} \tag{5.78}$$

$$v_k^{(f+1)} = \frac{\rho \left[\sigma^2 + \sum_{j \neq k} v_j^{(f)} \left| \mathbf{h}_{k,C_j^{(f)}}^H \tilde{\mathbf{g}}_j^{(f)} \right|^2 \right]}{\left| \mathbf{h}_{k,C_k^{(f)}}^H \tilde{\mathbf{g}}_k^{(f)} \right|^2}. \tag{5.79}$$

The downlink equal-rate methodology, corresponding to the gray box in Figure 5.25, is summarized below.

Downlink equal-rate algorithm

1. **Initialize**
   Set $f := 0$, start with no users in outage. Let $p_k^{(f)} = v_k^{(0)} = 0$ for all $k = 1, \ldots, KB$. For conventional networks, let $\mu_k^{(f-1)} = 0$.
2. For $f \geq 0$, given $\left\{ p_k^{(f)} \right\}$ and $\left\{ v_k^{(f)} \right\}$ (and $\left\{ \mu_k^{(f-1)} \right\}$), do the following:
   a. **Assign users**
      Assign each user ($k = 1, \ldots, KB$) to the base/cluster where it would attain the highest dual uplink SINR: (5.67) for conventional or (5.76) for network MIMO.
   b. **Compute dual uplink beamformers**
      Compute for each user ($k = 1, \ldots, KB$) the corresponding unit-norm beamforming weight derived from the MMSE criterion for the dual uplink: (5.68) for conventional or (5.77) for network MIMO.
      For conventional networks, update the interference power (5.69).
   c. **Update transmit powers**
      Update for each user ($k = 1, \ldots, KB$) the dual uplink powers $p_k^{(f+1)}$:
      (5.70) for conventional or (5.78) for network MIMO.
      Update for each user ($k = 1, \ldots, KB$) the downlink powers $v_k^{(f+1)}$:
      (5.71) for conventional or (5.79) for network MIMO.

d. **Determine users in outage**

For each base $b = 1, \ldots, B$, if $\sum_{k \in U_b} v_{k,b}^{(f+1)} > 1$, then eliminate users in decreasing order of requested power until the power constraint is met.

e. Let $f := f + 1$ and iterate until convergence.

5.5 Practical considerations

In this section, we discuss some practical aspects of implementing a cellular network, including the antenna array design, resource scheduling, and challenges for base station coordination.

5.5.1 Antenna array architectures

The antenna array architecture for a given sector is characterized by the number of antenna elements, the spatial correlation between the elements, and their directional response $G_{k,b}$ of the sector beam pattern. More sophisticated models account for the vertical beam pattern as well.

Figure 5.27 shows the physical manifestations of different array architectures used for each sector of a site with $S = 3$ sectors. Each rectangle denotes a radome enclosure and contains a column of either vertically polarized (V-pol) elements or cross-polarized (X-pol) elements. Within a V-pol column, the elements are co-phased to provide vertical aperture gain. Typically, the 3 dB beamwidth measured in the vertical direction is less than 10 degrees. Because the beamwidth is inversely proportional to the aperture size [154] and because the desired vertical beamwidth is narrower than the horizontal beamwidth (about 65 degrees for the case of 3-sector sites), the resulting antenna is longer in the vertical dimension. To help shape the beam pattern, a physical reflector is placed behind the array of elements to direct energy in the broadside direction. Note that the 1V configuration consists of multiple co-phased elements but for the purposes of performance evaluation is modeled by a single $M = 1$ antenna.

A commercial 1V antenna designed for a 2 GHz carrier frequency and $S = 3$ sectors per site (whose response is shown in Figure 5.6) has a height, width, and depth of, respectively, 130 cm, 15 cm, and 7 cm. The 3 dB beamwidth in the vertical and horizontal dimensions are 7 and 65 degrees, respectively. The dimensions are roughly proportional to the carrier wavelength, so an antenna with the same characteristics operating at 1 GHz would be twice as large in each dimension.

An X-pol antenna in Figure 5.27 consists of a column of dual-slant elements with ± 45 degree polarizations. The elements with different polarizations are orthogonal, so the X-pol antenna is modeled by $M = 2$ spatially uncorrelated antennas. The two antennas provide diversity, so the configuration is known as DIV-1X. Compared to the 1V antenna, the DIV-1X antenna is slightly wider. Diversity could also be achieved using two V-pol columns that have sufficient spatial separation (Section 2.4).

Base station antennas with half-wavelength spacing are highly correlated and can be used to form directional beams. The ULA-2V and ULA-4V configurations consist of, respectively, two and four columns of vertically polarized elements arranged in a uniform linear array (ULA) with half-wavelength spacing between columns. Because the aperture of ULA-4V is about twice that of the ULA-2V, it can form beams with about half the beamwidth.

The CLA-2X configuration consists of two X-pol columns with close (half-wavelength) spacing. One beam can be formed using the two columns of +45 degree polarization, and another beam can be formed in the same direction using the two columns of -45 degree polarization. Because the beams are formed using different polarizations, they will be uncorrelated. Therefore the CLA-2X configuration gives the benefits of both direction beamforming and diversity.

The DIV-2X configuration consists of two widely spaced columns to provide diversity. The spacing of the columns relative to the angle spread is sufficient so the columns are spatially uncorrelated. The DIV-2X configuration therefore achieves diversity order four. An array consisting of four V-pol columns could achieve the same diversity, but the array would be larger.

As discussed in Chapter 7, next-generation standards such as LTE-Advanced and IEEE 802.16m support MIMO techniques that use up to $M = 8$ antennas. Possible architectures include four closely-spaced X-pol columns, four widely spaced X-pol columns, or eight closely-space V-pol columns.

Physical form								
Antenna configuration	1V	ULA-2V	DIV-1X	TX-DIV	ULA-4V	CLA-2X	DIV-2X
Description	Single-column V-pol	Closely spaced V-pol columns	Single-column X-pol	Diversity spaced V-pol columns	Closely spaced V-pol columns	Closely spaced X-pol columns	Diversity spaced X-pol columns
Antennas per configuration	1		2			4	

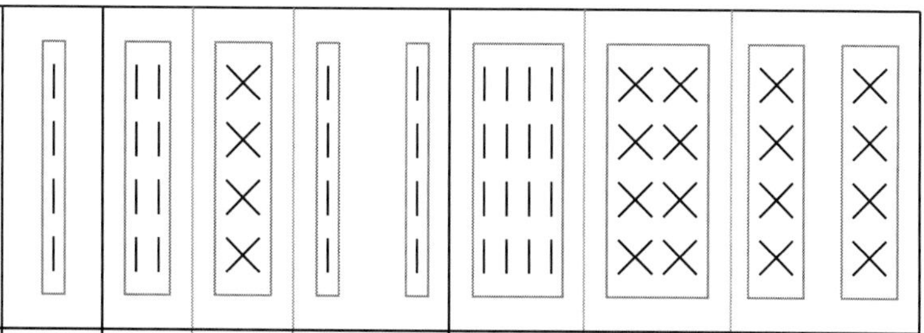

Fig. 5.27 Examples of antenna array architectures. Antenna elements in each column are co-phased to form a single virtual element.

5.5.2 Sectorization

For a site with $S = 3$ sectors and $M = 1$ antenna per sector, a 1V antenna can be used to create each sector. Using a ULA-2V array whose width is twice that of the 1V antenna, fixed beamforming weights implemented in the RF hardware can be applied to the two columns to create a beam whose width is half that of the 1V antenna's. Each ULA-2V array is enclosed in a single radome and is known as a *sector antenna*, and six ULA-2V arrays could be used to support a site with $S = 6$ sectors. Even though two columns of elements are used to create the sector, the weights across the elements are fixed, and there is effectively $M = 1$ antenna per sector. Similarly, twelve sector antennas, each implemented as a ULA-4V array, could support a site with $S = 12$ sectors and $M = 1$ antenna per sector.

Figure 5.28 shows the overhead view of different antenna configurations for cell sites. Columns A, B, and C show the site antenna configurations for $S = 3$, 6, and 12 sectors implemented with sector antennas. In doubling the number of sectors from $S = 3$ to 6 using sector antennas, there are twice as many antennas and each antenna doubles in width. Therefore the total weight of the antennas increases roughly by a factor of 4. The total weight again increases by a factor of 4 in going from $S = 6$ to 12 sectors. Larger antennas require additional infrastructure to cope with the higher weight and wind load, and they also are visually more obtrusive.

To support multiple $M > 1$ antennas per sector, the service provider could deploy M sector antennas in each sector. For example in column D of Figure 5.28, a pair of diversity-spaced V-pol columns (TX-DIV) is used for each sector of a $S = 3$-sector site to implement $M = 2$ uncorrelated antennas. Alternatively, in column E, a DIV-1X antenna could be used to implement $M = 2$ uncorrelated antennas with a smaller array footprint.

A linear array with closely spaced elements can be used as a sector antenna to form a single directional beam. It could also be used to form multiple simultaneous beams by weighting signals differently across the elements. The weighting could be performed in the RF domain through hardware (for example with Butler matrices) or electronically in the baseband domain. This implementation is sometimes known as a *multibeam antenna*. For example, as shown in column F of Figure 5.28, a ULA-2V array could create two beams, one pointing 30 degrees with respect to the broadside direction and another pointing at -30 degrees. Because the beams are not in the broadside direction, the sidelobe and beamwidth characteristics will not be as good as the one formed under column B's configuration. However, the antenna array footprint will be smaller. Alternatively, the ULA-2V configuration could be used to provide $M = 2$ correlated antennas for a sector of a $S = 3$-sector site. Column G of Figure 5.28 shows a site with three ULA-4V arrays. Each array could implement $M = 4$ correlated antennas in each of $S = 3$ sectors. Or each could create four fixed directional beams to support a total of $S = 12$ sectors. The characteristics of the directional beams can be improved by using additional elements, for example, using eight closely spaced V-pol columns instead of four. But the tradeoff is that the array is twice as wide. The beam characteristics and performance of these different antenna architectures will be discussed in Section 6.4.

Beams implemented in the RF hardware are fixed. Electronically generated beams provide more flexibility because they can be dynamically adjusted by changing the baseband beamforming weights. In channels with low angle spread, these dynamic beams could track the direction of mobiles using CDIT precoding (Section 4.2.2). The disadvantage of electronically generated beams is that the antenna elements need to be phase calibrated.

As discussed in Section 5.2.1, the effective sector response is widened as a result of channels with non-zero angle spread. Because the intersector interference increases as the sector response becomes wider, higher-order sector-ization is most effective if the angle spread is significantly smaller than the sector width. The angle spread associated with a base station tower decreases as the height of the tower increases relative to the height of the surrounding scatterers. For a fixed tower height, increasing the sectorization order will result in diminishing returns on throughput performance as the intersector interference overwhelms the multiplexing gains. In this regime, increasing the tower height could result in a lower angle spread and reduced intersector interference.

Angle spread distributions of a typical suburban channel and two urban macrocellular channels (derived from [36]) are shown in Figure 5.29. The suburban angle spread is less than 10 degrees with nearly probability 1. The angle spread of the urban channels is higher but is still less than 20 degrees with a significant probability. In all cases, dispersion due to angle spread would be negligible for $S = 3$ sectors. Suburban channels could support $S = 12$ channels without significant dispersion and urban channels could support fewer.

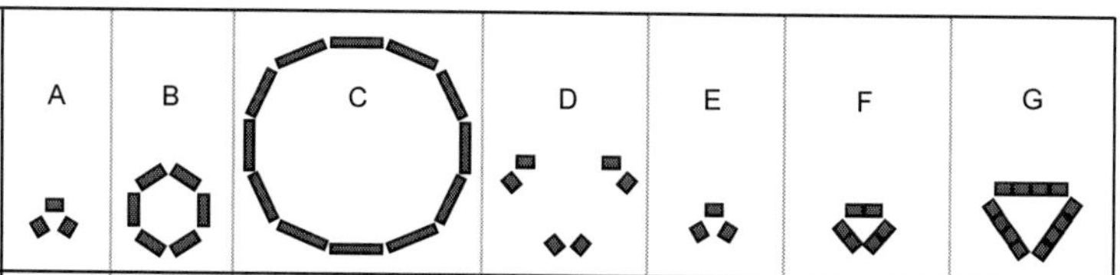

	A	B	C	D	E	F		G	
Overhead view									
Antenna configuration	ULA-1V	ULA-2V	ULA-4V	TX-DIV	DIV-1X	ULA-2V		ULA-4V	
Sectors per cell site (S)	3	6	12	3	3	3	6	3	12
Antennas per sector (M)	1	1	1	2UC	2UC	2C	1	4C	1

UC: spatially uncorrelated antennas
C: spatially correlated antennas

Fig. 5.28 Overhead view of a base station site showing antenna array cross-sections. The different antenna configurations are shown in Figure 5.27.

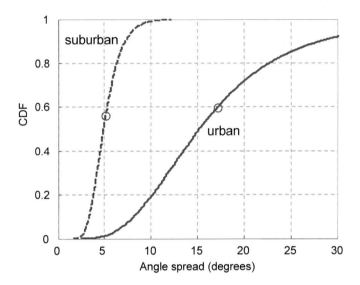

Fig. 5.29 Angle spread distribution of typical suburban and urban macrocellular channels [36]. The mean angle spread of the suburban and urban channels are, respectively, 5 and 17 degrees.

5.5.3 Signaling to support MIMO

In wireless systems such as cellular networks, auxiliary signaling is used to support MIMO transmission. For example, we have discussed in Sections 2.3.7 and 4.2.3 the feedback bits used to indicate a user's preferred codeword selected from the codebook $\mathcal{B}$. These B feedback bits are known as *precoding matrix indicator* (PMI) bits in the context of 3GPP LTE standards. Because the codewords are assumed to have unit power, the PMI conveys only directional information about the measured channel.

In order to set the coding and modulation for a data stream, an estimate of the channel magnitude, as measured by the SINR or geometry, is fed back from the mobile to the base. Due to the limited bandwidth on the uplink feedback channel, this estimate is quantized to a few bits. In 3GPP standards, they are known as *channel quality indicator* (CQI) bits. For a slowly fading SISO channel, reference signals associated with each base antenna can be used to estimate the channel h_{b^*} with respect to the serving base b^* and the channels $h_b, b \neq b^*$, of the interfering bases (Section 4.4.2). In this case, the CQI is a quantized measurement of the SINR

$$\frac{|h_{b^*}|^2 P_{b^*}}{\sigma^2 + \sum_{b \neq b^*} |h_b|^2 P_b}. \tag{5.80}$$

For MISO channels, common reference signals allow the estimation of the channels $\mathbf{h}_{b^*}^H$ and $\mathbf{h}_b^H, b \neq b^*$ (Section 4.4.2). If the desired base uses precoding vector $\mathbf{g}_{b^*}$, and the interfering bases use vectors $\mathbf{g}^b, b \neq b^*$, the SINR would be

$$\frac{\left|\mathbf{h}_{b^*}^H \mathbf{g}_{b^*}\right|^2 P_{b^*}}{\sigma^2 + \sum_{b \neq b^*} \left|\mathbf{h}_b^H \mathbf{g}_b\right|^2 P_b}. \tag{5.81}$$

However, since the precoding vectors of the interfering bases are not known, the user can approximate the interference power by assuming each base transmits isotropically. Then the CQI would be based on

$$\frac{\left|\mathbf{h}_{b^*}^H \mathbf{g}_{b^*}\right|^2 P_{b^*}}{\sigma^2 + \sum_{b \neq b^*} \|\mathbf{h}_b\|^2 P_b}. \tag{5.82}$$

For SU-MIMO transmission, spatially multiplexed streams could be encoded and modulated independently. In this case, CQI is fed back for each stream separately.

5.5.4 Scheduling for packet-switched networks

As described briefly in Chapter 1, a scheduler is used in packet-switched networks for dynamically allocating spectral resources among a base's assigned users. A block diagram of the scheduler and data transmission is shown in Figure 5.30. Scheduling is performed at each base on a frame-by-frame basis and is done independently for the uplink and downlink. Users are said to be *active* if they are scheduled to transmit data on the uplink or receive data on the downlink. The scheduler determines which users are active and how the frequency resources are allocated. If the base has $M > 1$ antennas, it determines how the spatial resources are allocated for spatial multiplexing. For each active user, the scheduler determines the precoding matrix, the number of streams transmitted (given by the rank of the precoding matrix), and the modulation and channel coding scheme for each stream.

For each user $k = 1, \ldots, K$, the scheduler requires the following information:

- performance metrics, such as the average achieved rate,
- QoS requirements, such as the maximum tolerated latency or minimum required rate,
- data statistics, such as the data buffer size,
- CSIT or PMI, if applicable, for precoding,
- CQI for determining the modulation and coding scheme.

The performance metrics, QoS requirements, and data statistics are readily available at the base. The PMI and CQI information is obtained through feedback from the mobiles, as described below in Section 5.5.5. Over a period of many frames, the scheduler seeks to optimize long-term objectives such as maximizing the mean throughput subject to user-specific rate and delay requirements. It can do so by appropriately updating the QoS weights as described in Section 5.4.2 and allocating resources to maximize the short-term weighted sum rate. We note that the base station ultimately decides how resources are allocated, and it is not obliged to follow the recommendations suggested by the CQI and PMI feedback. For example, even though a mobile feeds back a CQI to indicate it has a very high SINR, the base may schedule it a low rate transmission because there is very little data in the buffer for that user.

We have focused so far on the case of static, time-invariant channels. In practice, the channels experience time variations due to fading, and the channel-aware scheduler can exploit the variations. Figure 5.31 shows examples of two scheduling algorithms for a narrowband $(1, (2, 1))$ broadcast channel where a single-antenna base serves two single-antenna users over independently fading channels and where one active user is served per frame. As a result of the Rayleigh fading, the SINR of each user varies over time. The frame-by-frame SINR measured by each user is plotted versus time, and the scheduled user for the two algorithms is shown at the bottom of the figure. Round-robin scheduling is a simple algorithm which does not use the channel measurements. Each user is served on alternating frames regardless of the channel realizations. The channel-aware scheduler employs a simplified strategy for scheduling the user whose measured SINR is better than its average SINR. Over many frames, each user will be served half the time on average, and this strategy achieves a higher rate for each user than round-robin scheduling. As the number of users increases, the probability of a user

having a very good fade during any frame increases, and the advantage of
the channel-aware strategy is even higher compared to round-robin.

This effect of achieving improved system performance by exploiting independent fading across multiple users is known as *multiuser diversity*
[121, 155, 156]. In multiuser diversity, fading across users is exploited in order to improve the system performance, and larger fading variations actually
provide larger performance gains for multiuser diversity. Because transmit
diversity reduces the signal variations, it can actually result in decreased
throughput compared to a system with no diversity, when channel-aware
schedulers are employed. In fact, in the limit of a large number of users and
antennas, the maximum throughput achieved by *any* optimal scheduling algorithm can be infinitely worse than a system with no diversity [157].

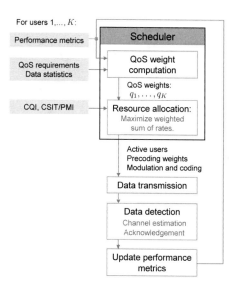

Fig. 5.30 Overview of scheduling and data transmission in a packet-switched cellular
network with scheduling. On a frame-by-frame basis, each base schedules and serves its
assigned users $1, \ldots, K$. Link adaptation is used to adjust rates based on the user's QoS
and channel information. The acquisition of CQI, CSIT, and PMI is described in Figure
5.32.

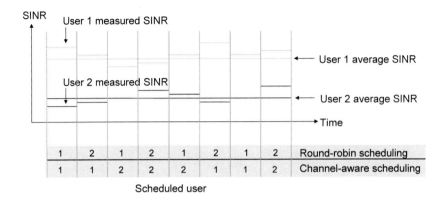

Fig. 5.31 Round-robin and channel-aware scheduling for a single-antenna base serving $K = 2$ users. The channel is assumed to be block fading.

5.5.5 Acquiring CQI, CSIT, and PMI

The steps for acquiring the CQI and CSIT/PMI information for scheduling are shown in Figure 5.32.

For the uplink, the base estimates the CSI of each active user based on their pilot signals and also the SINR using pilots from users assigned to other bases. Using the CSI and SINR estimates with the other user statistics available at the base, the scheduler determines the active users, data rates, and precoding matrices, if applicable. This information is conveyed to the users on a downlink control channel, and the active users transmit their data.

For the downlink, the sequence of events depends on whether the system is time-division duplexed (TDD) or frequency-division duplexed (FDD). In TDD systems, uplink and downlink transmissions occur over the same bandwidth but are duplexed in time. CSI estimates obtained from uplink user pilots could be used on the downlink, and the CSIT would be reliable if the channel changes slowly relative to the time-duplexing interval, as discussed in Section 4.4. The SINR of each user cannot be estimated from uplink transmissions and must be measured at the user. This information is conveyed back to the base on an uplink feedback control channel. With all the necessary information in hand, the base makes a scheduling decision and transmits to its active users.

In FDD systems, the SINR is estimated at the mobile and fed back to the base. CSI or PMI are likewise fed back on the uplink. For CDIT precoding, if the uplink and downlink channels are statistically correlated, the base could estimate the UL CDI and use this information for the downlink precoding.

The iterative scheduling algorithm described in Section 5.4.2 was described for time-invariant channels. The algorithm can be applied to time-varying channels by maximizing the weighted sum rate (5.27) using a rate vector $\mathbf{R}$ lying in the rate region $\mathcal{C}^{(f)}$ that varies as a function of the time index f. This algorithm forms the basis of channel-aware scheduling and can be extended for time-frequency resource allocation in wideband OFDMA systems [158]. The more general greedy-prime-dual algorithm [151] can also be used to dynamically allocate frequency resources in OFDMA systems [159].

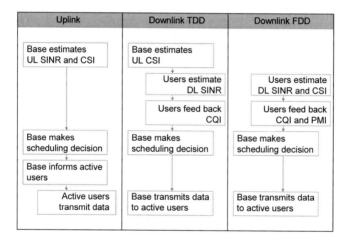

Fig. 5.32 Steps for acquiring CQI, CSI, and PMI information.

5.5.6 Coordinated base stations

Figure 5.33 shows architectures for implementing coordination base station techniques. Uplink network MIMO can be implemented using a centralized architecture as shown in the top subfigure for a coordination cluster of two bases. The architecture is centralized in the sense that the baseband processing for a cluster of coordinated bases is performed at a single location rather than in a distributed fashion across multiple bases. In the centralized architecture, each base is simply a remote radiohead which converts the radio-frequency (RF) signal to baseband. The baseband signal from each base (denoted as x_b for base $b = 1, 2$) is sent over the backhaul to a centralized baseband processor which jointly demodulates the signals. Estimates of the users' information streams, denoted by $\hat{u}_b$, are sent over another backhaul link to the core network. We note that the centralized baseband processing could be co-located with one of the radioheads, eliminating the need for backhaul to send its baseband signal.

Downlink coordination can also be implemented using a centralized architecture as shown in the center subfigure of Figure 5.33. Here, the centralized baseband processor generates the baseband signal s_b that is transmitted over base (radiohead) b. The information bit streams u_1 and u_2 are received from the core network, and the local CSIT (or baseband signals for estimating the CSIT) is obtained from each of the bases. The centralized processor therefore has global CSIT and can generate the baseband signal for the bases. For coordinated precoding, the transmitted signal from each base is a function of the data of its assigned user only. On the other hand, for network MIMO, the transmitted signal from each base is a function of the data of all users served by the coordination cluster.

Downlink coordination could be implemented using a distributed architecture, shown in the bottom subfigure of Figure 5.33, where each base performs its own baseband signal processing. This architecture is similar to the conventional architecture with independent bases, except that an enhanced backhaul is required. For coordinated precoding, each base receives the information streams of its assigned users, as in the conventional architecture. The enhanced backhaul is used for exchanging CSIT information between bases. Under the assumption of slowly varying pedestrian channels, the total backhaul requirement per base is only about 5% greater than the conventional backhaul requirement because the bandwidth required for updating the shared CSIT is minimal compared to the bandwidth of the data signals [160].

For network MIMO, the backhaul between the core network and the bases requires higher bandwidth because the transmitted signal from each base requires the data of both users. For coordinating a cluster of B bases, the backhaul bandwidth between the core network and each base increases by a factor of B to account for the user data. For small coordination clusters, the distributed architecture backhaul bandwidth is lower for the distributed architecture. For larger coordination clusters, the centralized architecture has lower backhaul requirements [160].

Downlink network MIMO transmission requires signals from multiple bases to arrive in a phase-aligned fashion at each user. Ideally, this requires tight synchronization so there is no carrier frequency offset (CFO) between the local oscillators at the base stations. Sufficient synchronization accuracy can be achieved using commercial global positioning system (GPS) satellite signals if the bases are located outdoors [161]. For indoor bases, the timing signal could be sent from an outdoor GPS receiver or via CFO estimation and feedback from mobiles [162].

5.6 Summary

In this chapter, we described the system model and simulation methodology for evaluating the performance of MIMO cellular networks.

- The physical layer of a cellular network can be characterized in terms of its system parameters (nature of the wireless channel, offered traffic patterns, etc.), design constraints (bandwidth, power, cost, etc.), and design outputs (base station and mobile requirements, air interface specification, etc.). Within this framework, the system engineer designs the appropriate antenna architectures, signaling strategies, and MIMO signal processing techniques to make best use of the spectral and spatial resources.
- Three aspects of the system model were highlighted: sectorization, reference SNR, and cell wraparound. Sectorization is the radial partitioning of a cell site, and its response as a function of azimuthal direction is modeled as a parabolic function. The reference SNR is a single parameter that conveniently encapsulates the effects of all parameters related to the link budget. Cell wraparound is a technique that addresses the reduced interference at the edge of a finite cellular network in order to ensure uniform performance statistics regardless of location.
- In conventional networks where base stations operate independently, intercell interference can be mitigated in the spatial domain either implicitly or explicitly using channel knowledge of the interferer. Coordinating the transmission and reception of base stations potentially reduces interference further at the expense of additional backhaul bandwidth.

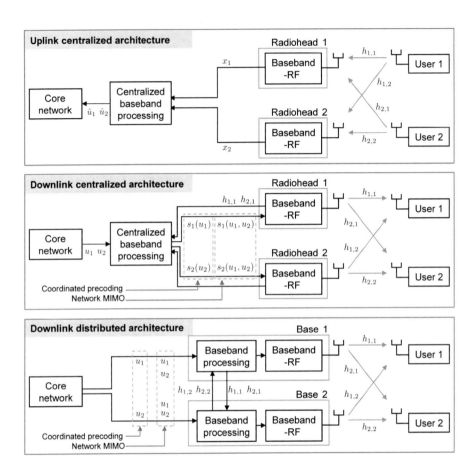

Fig. 5.33 Architectures for coordinated base stations. Uplink network MIMO can be implemented with a centralized architecture. Downlink coordination techniques (coordinated precoding and network MIMO) can be implemented with either a centralized or distributed architecture.

- Proportional-fair and equal-rate criteria are two strategies for operating a network to ensure fairness for all users. Uplink and downlink simulation methodologies are described for each criteria. Interference mitigation techniques using base station coordination, including coordinated precoding and network MIMO, are described for the equal-rate criteria.

Chapter 6
Cellular Networks: Performance and Design Strategies

In this chapter, we use the simulation methodology described in the previous chapter to evaluate the network performance of different MIMO architectures. We evaluate the performance of increasingly sophisticated systems, beginning with independent bases each with a single antenna, moving to independent bases with multiple antennas, and concluding with coordinated bases. In the final section of this chapter, we synthesize the conclusions for the various scenarios and outline a strategy for cellular network design under different environments.

6.1 Simulation assumptions

The assumptions for the simulations are described here, following the framework given in Section 5.1.

- **Channel characteristics**

  We assume a pathloss coefficient $\gamma = 3.76$, which is consistent with the Next-Generation Mobile Network (NGMN) Alliance's simulation methodology for modeling a typical macrocellular environment [135]. For interference-limited environments, the reference SNR is 30 dB. The shadow fading has standard deviation 8 dB which is a typical assumption for macrocellular networks [36] [135]. There is full shadowing correlation for co-located base antennas that serve different sectors and zero shadowing correlation for the users or bases otherwise. We assume the channel is frequency-nonselective with zero delay spread. The channel is static over a coding block, so the doppler spread is essentially zero. For most results, we assume rich scattering using an i.i.d. Rayleigh distribution. The exception is in Section 6.4.3 where the base antennas are totally correlated in order to implement direction beamforming. Unless otherwise noted, the angle spread at the base is zero to approximate the relatively small angle spreads found in macrocellular networks.

- **User statistics**
  Users are distributed uniformly throughout the network, and up to 30 users per site are considered. Mobile terminals are stationary or have low-mobility to be consistent with the block fading assumption. The data buffers for all users are always full.

- **Fundamental constraints**
  The power constraint is modeled as part of the reference SNR parameter. The carrier frequency affects the intercept of the pathloss model and is therefore accounted for by the reference SNR parameter.

- **Cost constraints**
  Because we are interested in establishing optimistic performance bounds, we do not constrain the computational complexity transceivers. Therefore we allow consideration of capacity-achieving DPC and MMSE-SIC techniques for the downlink and uplink performance, respectively. Further exploration of the relationship between performance and cost is beyond the scope of this book.

- **MAC and PHY-layer design**
  The air interface is assumed to be a packet-scheduled system, which is consistent with next-generation cellular standards. We model the performance of a single narrowband subcarrier of a wideband OFDMA air interface, and we assume the air interface is designed so there is no intracell interference. In other words, there is no interference between users served by a given base that are scheduled on different subcarriers.

- **Base station design**
  Bases are placed in a hexagonal grid of cells, and cell sites are partitioned into $S = 1, 3, 6, 12$ sectors. Unless otherwise noted, the sector response of the antennas is given by the model in (5.7). In scenarios with no base station coordination, the networks are shown in Figure 6.9. In scenarios with base station coordination, larger networks are used which are shown in Figure 5.26. In either case, cell wraparound (Section 5.2.3) is used to eliminate network boundary effects. Under the i.i.d. Rayleigh assumption, the base station antennas need to have sufficient spacing (greater than

10 wavelengths) or use dual-polarization. Under the correlated antenna model, the spacing is assumed to be a half wavelength.

- **Terminal design**

  The mobile antenna arrays are assumed to have an i.i.d. Rayleigh distribution. Because the mobiles are surrounded by local scatterers, this condition can be achieved with half-wavelength spacing of the elements.

The simulation assumptions and parameters are summarized in Figure 6.1. The proportional-fair criterion (with full-power downlink transmission and power-controlled uplink transmission) is used for networks with independent bases. The equal-rate criterion (with minimum sum power transmission) is used for networks with coordinated bases. When needed, ideal CSIT is assumed. This assumption is consistent with a TDD system with low-mobility users. Likewise CSIR is known exactly, a reasonable assumption for indoor applications [147] and for outdoor applications with slowly moving mobiles.

Distance-based pathloss	Pathloss coefficient: Intercept: Modeled by reference SNR
Shadow fading	8 dB standard deviation Fully correlated for co-located sectors Fully uncorrelated, otherwise
Fast fading	Time variation: Static, block fading Spatial correlation: i.i.d. Rayleigh Frequency variation: Flat
Channel state information (CSI)	Transmitter: Ideal, when applicable Receiver: ideal
Traffic model	Full buffer
Intrasector interference	None

Fig. 6.1 Parameters used for simulations in this chapter, unless otherwise noted.

6.2 Isolated cell

We first consider the SISO link performance in an isolated cell as a function of only the distance-based pathloss using an omni-directional antenna. From

(5.5), the spectral efficiency of a user distance d from the base is

$$C = \log_2 \left[1 + \frac{P}{\sigma^2} \left(\frac{d}{d_{\text{ref}}} \right)^{-\gamma} \right] \text{ bps/Hz}, \qquad (6.1)$$

where P/σ^2 is the reference SNR at distance d_{ref}. Writing the noise variance σ^2 explicitly in terms of the noise power spectral density N_0 and bandwidth W, we can express the achievable rate as the bandwidth multiplied by the spectral efficiency in (6.1):

$$R = W \log_2 \left[1 + \frac{P}{N_0 W} \left(\frac{d}{d_{\text{ref}}} \right)^{-\gamma} \right] \text{ bps}. \qquad (6.2)$$

Figure 6.2 shows the Shannon rate (6.2) as a function of the bandwidth W for a reference SNR of $P/\sigma^2 = 0$ dB given by the parameters for Downlink B in Figure 5.10. For very low and very high values of SNR, the following approximations are useful:

$$\log_2(1 + \text{SNR}) \approx \begin{cases} \text{SNR} \log_2 e, & \text{if SNR} \ll 1; \\ \log_2(\text{SNR}), & \text{if SNR} \gg 1. \end{cases} \qquad (6.3)$$

As the bandwidth decreases for a fixed power, the high-SNR approximation in (6.3) is valid and yields

$$R \approx W \log_2 \left[\frac{P}{N_0 W} \left(\frac{d}{d_{\text{ref}}} \right)^{-\gamma} \right]. \qquad (6.4)$$

If the bandwidth W is low, then increasing it results in a nearly proportional increase in the rate because the decreasing SNR inside the log term is insignificant compared to the linear term W outside the log. This region is known as the *bandwidth-limited region*.

If the bandwidth W is sufficiently high, the spectral efficiency becomes a linear function of the SNR. In this *power-limited* region, increasing the bandwidth does little to increase the rate, and eventually the rate reaches an asymptotic limit. Using the low-SNR approximation in (6.3), the limiting rate is

$$\lim_{W \to \infty} R = \frac{P}{N_0} \left(\frac{d}{d_{\text{ref}}} \right)^{-\gamma} \log_2 e.$$

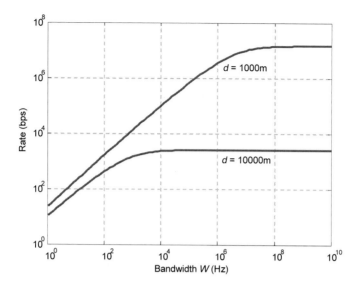

Fig. 6.2 Rate versus bandwidth for a given transmit power ($P/\sigma^2 = 0$ dB) and distance. As the bandwidth increases, the data rate saturates. As the distance increases, this saturation occurs at a lower bandwidth.

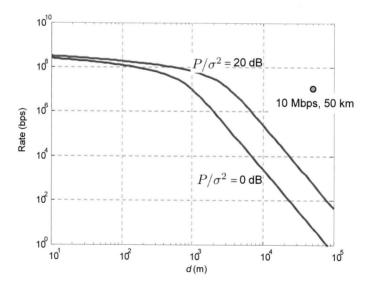

Fig. 6.3 Rate versus distance for fixed transmit power ($P/\sigma^2 = 0$ and 20 dB). For short distances, SNR is high and rate is proportional to log SNR. At large distances, SNR is low and rate is proportional to SNR. In order to achieve high data rates over a long distance, the transmit power must be very high. To achieve 10 Mbps at 50 km in 10 MHz bandwidth, the required transmit power is 400,000 W ($P/\sigma^2 = 67$ dB).

For the parameters of Downlink B in Figure 5.10, $P/\sigma^2 = 0$ dB and $W = 10$ MHz so that $P/N_0 = 10^7$ Hz. The limiting rates are 14.4 Mbps and 2.5 kbps, respectively for $d = 1000$ m and 10,000 m.

In the range between 1 MHz and 100 MHz where current and future cellular networks operate, the performance is somewhat power-limited at 1000 m and severely power-limited at 10,000 m. Using the parameters for Downlink B, 7.2 watts with a 20 dB building penetration loss is not enough power for additional bandwidth to increase the rate significantly.

Figure 6.3 shows the Shannon rate as a function of distance d for 10 MHz and infinite bandwidth. The performance is bandwidth-limited for low d and becomes power-limited as the receiver moves away from the transmitter. The curves highlight the fact that at sufficiently large distances from the transmitter, additional bandwidth will not increase the achievable rate significantly. In this power-limited regime, very high power is required to achieve high rates using SISO links. For example, to achieve 10 Mbps in 10 MHz at a distance of 50 km, a reference SNR of $P/\sigma^2 = 67$ dB is required, corresponding to 400,000 W transmit power. This magnitude underscores the difficulty of achieving high data rates at large distances using SISO links without violating the laws of physics or Shannon theory. Even with MIMO techniques, the size of the required power amplifiers is impractical, especially for uplink transmission.

Conclusion: Even with abundant spectrum, serving a geographic area with a single base is highly impractical due to the power requirements. This fact provides the motivation for using a cellular network.

6.3 Cellular network with independent single-antenna bases

In this section, we consider cellular networks where the bases have a single antenna, and study the effects of reduced frequency reuse, reduced cell size, and higher-order sectorization. As a baseline, we consider the performance of a network with universal frequency reuse and a single sector per site. The parameters for the four scenarios are summarized in Figure 6.4.

Sectors per site (S)	1	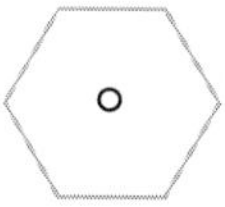 1	1	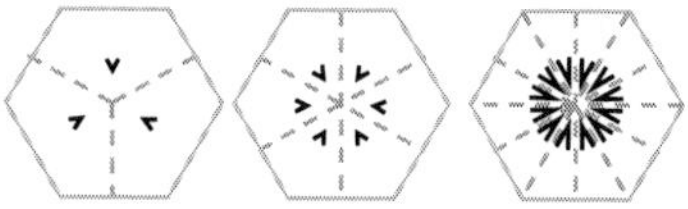 3	6	12
Antennas per sector (M)	1	1	1	1		
Transceiver	SISO	SISO	SISO	SISO		
Users per sector (K)	1	1	1	1		
Antennas per user (N)	1	1	1	1		
Reference SNR (P/σ^2)	-30 to 30 dB	10 dB	-30 to 30 dB	30 dB		
Frequency reuse (F)	1	1	1, 1/3, 1/7	1		
Transmission mode	Full-power	Full-power	Full-power	Full-power		
Section	6.3.1	6.3.2	6.3.3	6.3.4		

Fig. 6.4 Parameters for downlink system simulations that assume independent, single-antenna bases (Section 6.3).

6.3.1 Universal frequency reuse, single sector per site

We first study the downlink performance of a cellular network with universal frequency reuse where a single omni-directional antenna is used at each site. There is therefore one sector per site ($S = 1$). We assume that the channel realization is dependent on only the distance-based pathloss and shadowing and that there is no Rayleigh fading. A single user is assigned to each base ($K = 1$), and as a special case of the proportional-fair methodology for $K = 1$, we assume each base transmits with full power P. As a result, the geometry (5.36) for a given user does not depend on the location of other users served by other bases.

For very low transmit powers (low reference SNR), the normalized noise power in the denominator of (5.36) dominates over the interference term:

$$(P/\sigma^2) \sum_{b \neq b^*} \alpha_{k,b}^2 \ll 1.$$

In this regime, the system is *noise-limited*, and increasing P/σ^2 leads to a significant increase in the geometry. For high reference SNRs where the interference term dominates over the noise

$$(P/\sigma^2) \sum_{b \neq b^*} \alpha_{k,b}^2 \gg 1,$$

the system is *interference-limited*, and increasing P/σ^2 results in a negligible increase in the geometry.

For a given reference SNR P/σ^2, the geometry is a random variable due to the random placement of each user and the random shadowing. The cumulative distribution function (CDF) of the geometry is shown in Figure 6.5 for $P/\sigma^2 = -10, 0, 10$ dB. In the range from $P/\sigma^2 = -10$ dB to 0 dB, the system is noise limited so that increasing the transmit power results in a significant shift in the geometry CDFs. In the range from $P/\sigma^2 = 0$ dB to 10 dB, the system becomes interference limited so that the shift in the geometry curves is much less.

For a given realization of the geometry Γ, the achievable rate is $\log_2(1+\Gamma)$. Because there is one user assigned per site, the user rate is equivalent to the base throughput. Figure 6.6 shows the resulting CDF of the throughput, where the mean for each curve is indicated by the circle. The mean throughput is plotted versus the reference SNR in Figure 6.7. For reference SNRs greater than 10 dB, the performance is interference limited.

The reference SNR at which performance becomes interference limited depends on the relationship between the interference and noise power. Here we have assumed a single antenna per mobile. If the mobile had multiple antennas, then spatial processing would reduce the interference power so that the system would become interference limited at a higher reference SNR.

Conclusion: As the transmission power of all bases increases in a system with independent bases, the intercell interference power begins to dominate over the noise power, and the performance becomes interference limited. The maximum transmission power should be designed so the performance is operating in the interference-limited regime because otherwise, spectral efficiency could be improved by increasing the transmit powers.

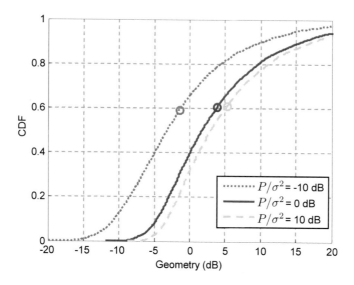

Fig. 6.5 CDF of downlink geometry (5.36) under universal frequency reuse. Means are indicated by the circles. As the reference SNR P/σ^2 increases, the geometry becomes interference limited. A 10-time increase in transmit power from $P/\sigma^2 = -10$ dB to 0 dB results in a significant shift in the CDF. From $P/\sigma^2 = 0$ dB to 10 dB, the shift is much less.

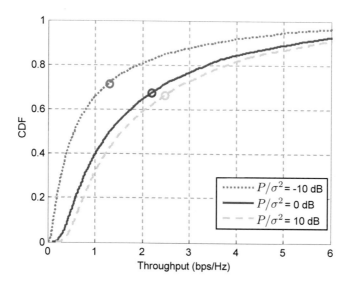

Fig. 6.6 CDF of downlink throughput under universal frequency reuse. Throughput is based on the geometry with no Rayleigh fading.

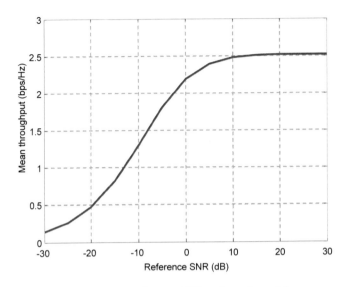

Fig. 6.7 Mean throughput versus reference SNR under universal frequency reuse.

6.3.2 Throughput per area versus cell size

Suppose that a given reference SNR P/σ^2 is achieved in a network with cells of radius r_1 (Figure 5.9) using transmit power P_1. For a network with cells of radius $r_2 < r_1$, the transmit power P_2 required to achieved the same reference SNR, assuming all other parameters are fixed, is obtained by solving

$$P_1 r_1^{-\gamma} = P_2 r_2^{-\gamma}, \tag{6.5}$$

where γ is the pathloss exponent. Hence for a fixed reference SNR, the required transmit power for a system with cell radius r_2 is reduced by a factor of $(r_1/r_2)^{\gamma}$. (With $\gamma = 3.76$, a cell with radius $r_2 = 0.5r_1$ requires about a factor of 14 less power.) If the number of users per cell is fixed and independent of the cell radius, reducing the cell size increases the throughput per unit area by a factor of $(r_1/r_2)^2$ because the number of bases per unit area scales with the cell radius squared.

Furthermore, we note that the scaling factor for the transmit power *per unit area* is $(r_1/r_2)^{\gamma-2}$. Hence for $\gamma > 2$, smaller cells increase the throughput

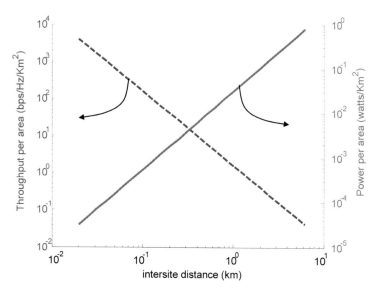

Fig. 6.8 The throughput per area (dashed line) and transmit power per area (solid line) are plotted as a function of intersite distance for a cellular network with universal reuse and one sector per cell. The reference SNR is fixed at $P/\sigma^2 = 30$ dB. As the cell size decreases, the throughput per area increases and the transmit power per area decreases. Therefore smaller cells result in higher total throughput using lower total transmit power.

per area while requiring less total power per area. This observation provides strong motivation for using small cells to serve densely populated areas with high service demands. In practice, these small cells, sometimes known as femtocells or picocells, should be deployed when the demand for service justifies the total costs that include backhaul and baseband processing. These small cells are especially effective when used indoors to provide indoor coverage because the exterior building walls provide significant attenuation to reduce interference between the indoor and outdoor networks.

Figure 6.8 shows the throughput per area and the total power per area as a function of the intersite distance. At a reference distance of 1 km (corresponding to an intersite distance of 2 km), a reference SNR of 10 dB can be achieved using a transmit power of 0.72 watts, assuming 0 dB penetration loss and with all other parameters given by the parameters for Downlink B in Figure 5.10. From Figure 6.12, the mean throughput is 2.5 bps/Hz. The area of a cell with radius 1 km is 6.9 km^2. At an intersite distance of 2 km, the throughput per area is 0.36 bps/Hz per km^2, and the power per area is 0.10 watts per km^2. Decreasing the intersite distance by a factor of 10 to 100 m, the 10 dB reference SNR (and hence the same mean throughput per site) can be achieved by transmitting with a factor $10^3.76 \approx 5700$ less power per site. The density of sites increases by a factor of 100. Therefore the throughput per area increases by a factor of 100 while the power per area decreases by a factor of 57.

We emphasize that these scaling results assume that the number of users per cell is fixed, independent of the cell size. If the number of users for the total area of the network is fixed, then as the cell size decreases, the average throughput would decrease significantly as some cells have no users to serve. The scaling results assume that the pathloss coefficient is also independent of the cell size. In practice, as the cell size decreases and the scattering channel becomes more like a line-of-sight channel, then the coefficient decreases. For example, the pathloss coefficient for a line-of-sight microcell channel is 2.6 [36]. In this case, the additional interference would shift the geometry distribution curve to the left, and the throughput per cell would decrease.

For the uplink, if the user density and pathloss coefficient are independent of the cell size, then the uplink throughput per cell is also independent of the cell size, and the same scaling results apply.

Conclusion: If the number of users per cell and the pathloss coefficient are independent of the cell radius, then as the radius decreases, the throughput per area increases and the total transmitted power per area decreases. In practice, the cost per site, including expenses for infrastructure, backhaul connections, and site leasing, need to be taken into account when determining the cell site density.

6.3.3 Reduced frequency reuse performance

Sections 6.3.1 and 6.3.2 assumed universal frequency reuse where all B bases transmit on the same frequency. Under reduced frequency reuse (Section 1.2.1), adjacent cells operate on different subbands, resulting in reduced intercell interference. We study the performance of reduced frequency reuse where the bandwidth W is partitioned into multiple subbands of equal bandwidth. The parameter $F \leq 1$ is known as the *frequency reuse factor*, and we consider the cases of universal reuse ($F = 1$), reuse $F = 1/3$ with 3 subbands, and reuse $F = 1/7$ with 7 subbands. To minimize co-channel interference, the subbands are assigned using a reuse patterns shown in Figure 6.9.

For a given reuse factor F, we assume that each base transmits with power P over bandwidth FW. If the noise spectral density is N_0, then the noise variance will be $F\sigma^2 = FN_0W$. If we generalize the expression for the geometry (5.36), the noise variance is reduced by a factor of F, and the geometry of a user assigned to base b^* is given by

$$\frac{(P/\sigma^2)\alpha_{b^*}^2}{F + \sum_{b \neq b^*}(P/\sigma^2)\alpha_b^2}, \tag{6.6}$$

where the summation occurs over bases operating on the same subband as base b^* and where we have simplified the notation by removing the user index.

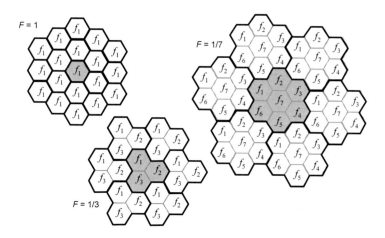

Fig. 6.9 Frequency assignment for cells under frequency reuse F. Under universal frequency reuse ($F = 1$), all cells use the same frequency. Under reuse $F = 1/3$, the bandwidth is partitioned into 3 bands f_1, f_2, f_3. Under reuse $F = 1/7$, the bandwidth is partitioned into 7 bands $f_1, \ldots, f_7$.

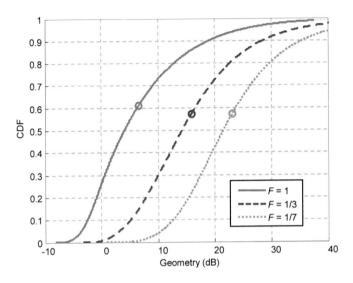

Fig. 6.10 CDF of downlink geometry (6.6), parameterized by frequency reuse F, for reference SNR $P/\sigma^2 = 30$ dB. Means are indicated by the circles. Reducing frequency reuse results in less interference and higher geometry.

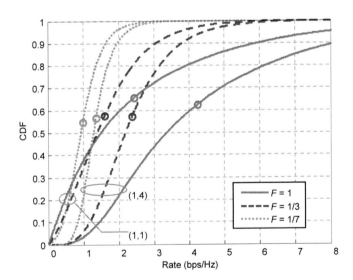

Fig. 6.11 CDF of downlink rate/throughput (6.9), parameterized by frequency reuse F, for reference SNR $P/\sigma^2 = 30$ dB. Means are indicated by the circles. Universal reuse achieves the highest mean throughput. Under SISO, reduced reuse achieves higher cell-edge rates, but with multiple receive antennas, universal reuse achieves highest cell-edge rate.

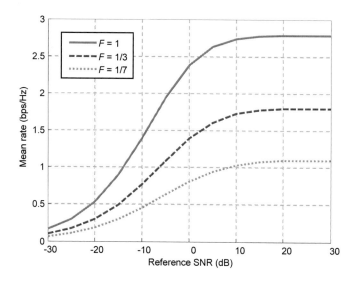

Fig. 6.12 Mean rate versus reference SNR for (1,1) links, parameterized by frequency reuse F. The mean rate performance is interference-limited for $P/\sigma^2 \geq 10$ dB for all frequency reuse factors F.

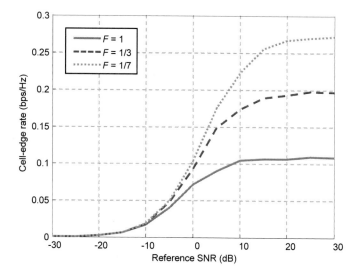

Fig. 6.13 Cell-edge rate versus reference SNR for (1,1) links, parameterized by frequency reuse F. The cell-edge rate performance is interference-limited for $F = 1$ for $P/\sigma^2 \geq 10$ dB. For reduced reuse, the cell-edge rate becomes interference limited at higher reference SNR.

Figure 6.10 shows the distribution of the geometry (6.6) for a reference SNR of $P/\sigma^2 = 30$ dB and for $F = 1, 1/3$, and $1/7$. As F decreases, the interference power decreases for a given user location, and as a result, the geometry curves shift to the right.

If we let $\mathbf{h}_b \in \mathbb{C}^N$ be the channel realization from base b to a user with N antennas, the output of a maximal ratio combiner based on the desired base's channel $\mathbf{h}_{b^*}$ is

$$\mathbf{h}_{b^*}^H \mathbf{x} = |\mathbf{h}_{b^*}|^2 s_{b^*} + \sum_{b \neq b^*} \mathbf{h}_{b^*}^H \mathbf{h}_b + \mathbf{h}_{b^*}^H \mathbf{n}, \tag{6.7}$$

and the resulting SINR is

$$\frac{P/\sigma^2 ||\mathbf{h}_{b^*}||^2}{F + \sum_{b \neq b^*} P/\sigma^2 \frac{|\mathbf{h}_{b^*}^H \mathbf{h}_b|^2}{||\mathbf{h}_{b^*}||^2}}. \tag{6.8}$$

Because each base uses only a fraction F of the total bandwidth, the spectral efficiency is normalized by F and is given by

$$F \log_2 \left(1 + \frac{P/\sigma^2 ||\mathbf{h}_{b^*}||^2}{F + \sum_{b \neq b^*} P/\sigma^2 \frac{|\mathbf{h}_{b^*}^H \mathbf{h}_b|^2}{||\mathbf{h}_{b^*}||^2}}\right). \tag{6.9}$$

Multiple antennas at the mobile ($N > 1$) provide combining gain, manifested by the term $||\mathbf{h}_{b^*}||^2$ in the numerator, and interference-oblivious mitigation (Section 5.3.1), manifested by the term $|\mathbf{h}_{b^*}^H \mathbf{h}_b|^2$ in the denominator.

Figure 6.11 shows the CDF of rates generated from (6.9) for reference SNR $P/\sigma^2 = 30$ dB using Rayleigh channels with $N = 1$ and 4. While reduced reuse has an advantage over $F = 1$ for the SINR (6.8), this advantage occurs inside the log term of (6.9), whereas the factor F outside the log term results in a linear reduction of the rate. While universal reuse has the lowest geometry, its rate performance is often better than the reduced reuse options because of the relative advantage of the scaling term $F = 1$ outside the log. In particular, the mean throughput is highest using universal reuse for both $N = 1$ and $N = 4$.

For $N = 1$, universal reuse has the highest peak (90% outage) rate but $F = 1/7$ reuse has the highest cell-edge (10% outage) rate. If multiple receive antennas are used, the SINR (6.8) benefits from combining gain of the desired signal and spatial mitigation of the interference. For $N = 4$, universal reuse achieves the highest peak rate. For the cell-edge performance, because the geometry of universal reuse (-2 dB) is in low-SNR regime of (6.3), a linear increase in the SINR results in a linear increase in rate. On the other hand, the cell-edge geometry of reuse $F = 1/7$ (13 dB) is in the high-SNR regime so that a linear increase in SINR results in only a logarithmic improvement in rate. Therefore the relative improvement in cell-edge rate due to combining provides a more significant gain for $F = 1$, and its cell-edge rate is the best. In an isolated link at low SNR, MRC combining results in a rate gain of about 4 (see Figure 2.7). However, in a system context, MRC provides additional gain due to interference mitigation, so the total gain in cell-edge rate for $F = 1$ is about a factor of 10.

For $N = 1$, the mean rate and cell-edge rate performance versus the reference SNR are shown respectively in Figures 6.12 and 6.13. The mean rate performance is interference limited for all three values of F beyond 10 dB. However, the cell-edge rate becomes interference limited at a higher reference SNR when $F < 1$ because the noise power is still relatively large compared to the interference power at the cell edge.

Note that if each cell in the network is assigned a unique frequency band, the throughput performance of each cell would be noise limited. Therefore an arbitrarily high throughput could be achieved by increasing the transmit power. However, in a system with B bases, the bandwidth per base is W/B, and the transmit power required to achieve a reasonable data rate would be impractically high for any network having more than a few cells.

Conclusion: The mean throughput of a typical cellular network is maximized under universal frequency reuse where each cell uses the entire channel bandwidth. Reduced frequency reuse has the advantage of better cell-edge rate performance if the mobiles have only a single antenna, but this advantage is reduced with multiple-antenna mobiles.

6.3.4 Sectorization

Let us consider a downlink cellular network with S sectors per site and $M = 1$ antenna per sector. The sectors are assumed to be ideal for now and the sector antenna response is given by (5.6). Universal reuse is assumed so that the same frequency resources are reused in all sectors and cell sites. We fix the total transmitted power per site so the power transmitted in each sector is $1/S$ the total power. However, as a result of the sector antenna gain, the received power from any base is independent of S. As a result, the statistics of the channel variance α_b^2 for the desired base sector is independent of the number of sectors S. Likewise, the channel variance for any interfering sector is independent of S. Therefore, because a user in a given location receives non-zero power from exactly 1 sector per site as a result of the ideal sector response, the resulting geometry distribution is independent of S. Assuming that the channel statistics of each sector are identical, it follows that for a given reference SNR P/σ^2, the throughput distribution *per sector* is independent of S. Therefore a network with S sectors per site with ideal sectorization has a mean throughput S times that of a system with omnidirectional base antennas.

Assuming i.i.d. Rayleigh channels for the users, then as the number of users increases asymptotically, the sum rate per sector scales with $\log \log K$ (Section 3.6). If we use a single panel antenna per sector, then with M sectors per site, the sum rate per site scales with $M \log \log K$. In this sense, ideal sectorization achieves the optimal sum-rate scaling of a cell site with M transmit antennas. Compared to other scaling-optimal strategies such as DPC or ZF beamforming, sectorization is much simpler to implement because it transmits with only a single antenna in each sector. As a result, it requires neither CSI at the transmitter nor computation of beamforming weights.

Similar throughput scaling can be achieved using non-ideal parabolic antenna responses (5.7) if the intersector interference does not increase as S increases. This condition can be met if we use the sector antenna parameters in Figure 5.5. Figure 6.14 shows the resulting CDF of the geometry calculated using (5.36) for users distributed uniformly in a cellular network with $S = 1, 3, 6, 12$ sectors under a reference SNR of $P/\sigma^2 = 30$ dB. The geometry distribution for $S > 1$ sectors per site is slightly degraded from the $S = 1$ geometry as a result of intersector interference. Because the shadowing realization is identical for sectors associated with a given site, the peak geometry occurs when a user is very close to a base and located in the boresight direction of the serving sector. If the received power from the serving sector is P', then, using the parameters in Figure 5.5, the power received from an interfering sector co-located with the serving sector is $\frac{P'}{(3/S)100}$ for $S = 3, 6, 12$. Therefore because there are $S - 1$ interfering sectors per site, the peak geometry is $\frac{100S}{3(S-1)}$. As seen in Figure 6.14, the peak geometry decreases a small amount in going from $S = 3$ to 6 to 12 sectors. The peak geometry for $S = 1$ is much higher because there is no co-site intersector interference.

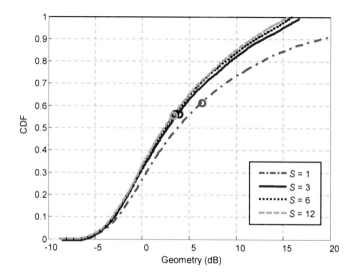

Fig. 6.14 CDF of geometry (5.36), parameterized by S, the number of co-channel sectors per site. For $S \geq 3$, the geometry is limited as a result of interference from other sectors belonging to the same site. As a result of the sector response parameters in Figure 5.5, the interference characteristics for $S = 3, 6, 12$ are similar.

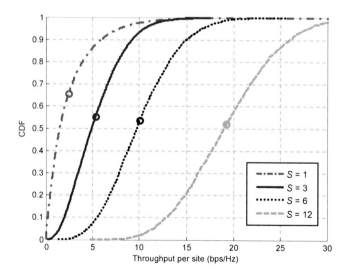

Fig. 6.15 CDF of throughput per site based on the geometries from Figure 6.14. Mean throughput per site approximately doubles in going from $S = 3$ to 6 sectors and again from $S = 6$ to 12 sectors.

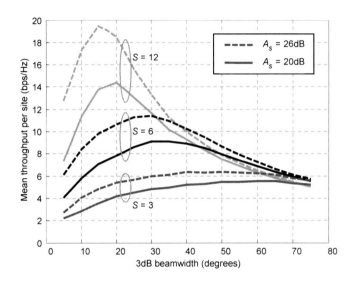

Fig. 6.16 Mean throughput per site versus sector beamwidth for $S = 3, 6, 12$ sectors per site, parameterized by the sidelobe levels A_s. Zero-degree angle spread is assumed. For the sidelobe levels given in Figure 5.5, the throughput performance is nearly optimal for the corresponding beamwidths.

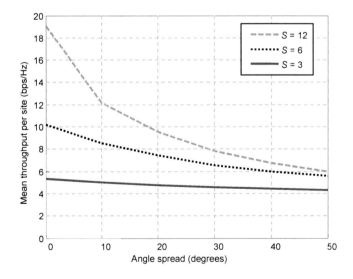

Fig. 6.17 Mean throughput per site versus RMS angle spread, using the sector parameters in Figure 5.5. As the angle spread increases, the effective sector response becomes wider, resulting in additional interference and degraded performance. For a given angle spread, performance degradation for narrower sectors is more severe.

With regard to the overall geometry distribution, the CDF curve for $S = 3$ is similar in shape to that for $S = 1$ but shifted to the left as a result of additional interference. For $S = 3$, 6 and 12, because the beamwidth $\Theta_{3\text{ dB}}$ and the sidelobe level A_s parameters are scaled by a half for each doubling of S, the distribution of the geometry for a user k at a given angle $|\theta_{k,b^*} - \omega_{b^*}| \leq \pi/S$ with respect to its serving sector b^* under $S = 3$ will be similar to the distribution at the angle $(\theta_{k,b^*} - \omega_{b^*})/2$ under $S = 6$ and at the angle $(\theta_{k,b^*} - \omega_{b^*})/4$ under $S = 12$. Therefore if the users are uniformly distributed, the CDF of the geometry will be similar for $S = 3, 6, 12$.

Because the geometry distributions for $S = 3$, 6, 12 are similar, the resulting throughput distributions *per sector* are also similar. A throughput realization *per site* is obtained by summing the throughput across S co-located sectors for a given placement of users and channel realizations, and the resulting CDF is shown in Figure 6.15. As a result of the central limit theorem and the fact that the throughput realizations per sector are independent, the throughput distribution per site becomes more Gaussian as S increases. The mean throughput roughly doubles in going from $S = 1$ to 3. Because the throughput per sector is largely independent of S (for $S \geq 3$), the mean throughput per site doubles in going from $S = 3$ to 6, and from $S = 6$ to 12. Therefore, as a result of the sector response parameters in Figure 5.5, the intersector interference is independent of S, and the mean throughput under higher-order sectorization with parabolic antenna patterns scales linearly with $S \geq 3$, as was observed in the ideal sectorization case.

What happens to the throughput if the sector beamwidth is wider or narrower than the ones defined in Figure 5.5? Figure 6.16 shows the performance mean throughput performance, still using parabolic sector responses (5.7), but allowing the sector beamwidth (measured by the parameter $\Theta_{3\text{ dB}}$) to vary from 5 to 75 degrees for $S = 3, 6, 12$. The sidelobe parameter A_s is either 20 dB or 26 dB. This figure highlights the importance of designing the sector response to match the number of sectors S. For $S = 3$ and $A_s = 20$ dB, the throughput is essentially flat in the range of $\Theta_{3\text{ dB}}$ from 50 to 70 degrees. Therefore using a beamwidth of 70 degrees is reasonable. For $S = 12$ and $A_s = 26$ dB, using a beamwidth of 18 degrees nearly maximizes the throughput. The performance of higher-order sectorization, especially for $S = 12$, is sensitive to the sidelobe levels and beamwidth.

The simulation results in Figures 6.15 and 6.16 assume that the channel has zero angle spread. For a non-zero angle spread, the effective sector pattern can be computed as described in Section 5.2.1. As the angle spread increases, the effective sector beamwidth increases (see Figure 5.8), resulting in additional intercell and intersector interference. Figure 6.17 shows the impact of higher angle spread on the throughput performance of sectorization. The performance of higher-order sectorization is more sensitive because for a fixed angle spread, there is more interference. In going from 0 to 10 degrees angle spread, the throughput for $S = 12$ is reduced by almost 40% whereas the throughput for $S = 6$ is reduced by only 20%. However, the absolute throughput for $S = 12$ is still superior and is over twice the throughput of conventional sectorization $S = 3$. As the angle spread increases beyond 40 degrees, there is very little throughput advantage in using $S = 12$ compared to $S = 6$ sectors.

While sectorization is a relatively simple way of increasing throughput, a few factors limit the sectorization order, preventing a designer from dividing a cell into an arbitrarily large number of sectors. First, the azimuthal beamwidth of the sector should be sufficiently large compared to the angle spread of the channel so that the sidelobe energy falling into adjacent sectors is tolerable. As a result, the geometry distribution will be independent of S. Second, the physical size of the antenna should meet aesthetic and wind load requirements. Third, if users are not uniformly distributed and there are too few users in a sector, then higher-order sectorization would not be justified. Fourth, the handoff rate may become too high if there are too many sectors. Within these limits, higher-order sectorization is an effective way to increase cell throughput and should be used as a baseline for comparison against more sophisticated MIMO techniques. Recall that electronic beams have several advantages over a panel implementation to address these factors, including a smaller physical size and flexibility to steer beams and partition power among them.

Conclusion: Sectorization achieves multiplexing gain for a cell site by serving multiple users over different spatial beams. Ideal sectorization (with no interbeam interference) achieves the optimal sum-rate scaling per site. Using conventional, non-ideal sectorization with $S = 3$ as a baseline, the throughput with higher-order sectorization scales in direct proportion to S if the geometry distribution for a sector is independent of S and if the channel angle spread is significantly less than the sector beamwidth. If the angle spread is too large, interbeam interference diminishes the gains.

6.4 Cellular network with independent multiple-antenna bases

Using multiple antennas, the bases can employ SU- and MU-MIMO techniques to improve the system performance. Treating interference as noise, we show that the area spectral efficiency exhibits the same linear scaling with respect to the number of antennas as in isolated links. For a fixed number of antennas *per site*, we study the tradeoff between implementing fewer sectors per site with more antennas per sector and implementing more sectors per site with fewer antennas per sector.

6.4.1 Throughput scaling

The capacity gain for MIMO channels compared to SISO channels was first discussed in Chapter 1. The spatial multiplexing gains, which are defined for asymptotically high SNRs, are observed for a wide range of SNRs. Figure 1.8 shows these gains can be achieved even for moderate SNRs, and Figure 1.9 shows that similar gains can be achieved in the low-SNR regime as a result of combining. How are these gains reflected in a cellular system where the performance is affected by interference and where the geometry of the users is different?

Figure 6.19 shows the CDF of the throughput per site for $S = 3$ sectors per site under ideal sectorization (5.6) and non-ideal sectorization (5.7) using parameters from Figure 5.5. The performance is shown for the case of single-user SISO transmission ($M = 1$, $K = 1$, $N = 1$), single-user closed-loop MIMO transmission ($M = 4$, $K = 1$, $N = 4$), and sum-capacity multiuser MISO transmission ($M = 4$, $K = 4$, $N = 1$). (A maximum sum-rate criterion is used for MU-MISO.) The channels are assumed to be i.i.d. Rayleigh. These parameters are summarized in Figure 6.18.

Under ideal sectorization, the distribution of the throughput per sector is independent of the sectorization order S. Therefore the SU-SISO throughput per site for $S = 3$ is a random variable equivalent to the summation of three independent realizations of the SU-SISO throughput for $S = 1$ per site given in Figure 6.15. As a result, the mean throughput for $S = 3$ is three times the mean throughput for $S = 1$.

In going from single-user SISO transmission to closed-loop MIMO transmission, the throughput increases as a result of combining gain for users with lower geometry and spatial multiplexing for users with higher geometry. At lower geometries (-10 dB to 0 dB), the MIMO system throughput gain (where the interference is modeled as being spatially white, as in equation (5.38)) can be inferred from Figure 2.7 for SNRs in the range of -10 dB to 0 dB. Due to the inequality in (5.43), the mean throughput obtained when modeling the Rayleigh fading and colored interference is higher. Overall, the gain at lower geometries is greater than 4 as a result of combining gain. For example, at the cell edge (10% outage) throughput gain is about a factor of 6.8 in going from 2.5 bps/Hz to 17 bps/Hz. For users with higher geometries, the MIMO throughput gain is about a factor of 4 due to spatial multiplexing. (The peak throughput at 90% outage increases from 14 bps/Hz t 56 bps/Hz for a gain of 4.) Overall, the mean throughput increases by about a factor of 4.5, from 7.8 bps/Hz to 35 bps/Hz. (If we modeled the interference as spatially white (5.38), then the mean throughput increases by a factor of 4.3, from 7.2 bps/Hz to 31 bps/Hz. This result is not shown in Figure 6.19.)

Under MU-MISO transmission with ideal sectorization, multiple users are distributed across the cell and geometry realizations drawn from the CDF in Figure 6.14 for $S = 1$. As a result of multiuser diversity, the throughput distribution would have less variance than the case where the multiple users have the same realization. By coordinating the reception of users located in the same location, the throughput would be improved, and this case would correspond to the SU-MIMO curve in Figure 6.19. Therefore in going from SU-MIMO transmission to MU-MISO transmission, the CDF becomes steeper and the mean throughput decreases.

Compared to the performance with ideal sectorization, the performance using the non-ideal sector response is degraded as a result of intersector interference. Because the degradation in the geometry is more significant at the upper tails as shown in Figure 6.14, the degradation in throughput is likewise more significant at the upper tails of its CDF.

The mean throughput for the case of nonideal parabolic sector responses (5.7) is plotted in Figure 6.20 as a function of M for $M = 1, 2, 4$. (For $M = 2$, SU-MIMO transmission assumes $K = 1$ and $N = 2$, and MU-MIMO transmission assumes $K = 2$ and $N = 1$.) In addition to the performance under the maximum sum-rate criterion for the downlink, corresponding site throughput performance is also shown for the uplink and for both the downlink and uplink under the equal-rate criterion. The linear scaling with respect to M is apparent from this figure. Under the equal-rate criteria, the throughput per site is lower because the bases transmit with less than full power. Despite the differences in the absolute multiuser throughput between the two criteria, the scaling gains are very similar. Under the proportional-fair criterion, the multiuser downlink performance is reduced compared to the maximum sum-rate criterion. For $M = 4$, $K = 4$, $N = 1$, the mean throughput per site is 19 bps/Hz, and the throughput gain with respect to SU-SISO is 3.8. These results are shown later in Figure 6.26 but not in Figure 6.20.

Conclusion: When MIMO techniques are implemented in a cellular network, the average throughput increases linearly with respect to the number of antennas. These gains are achieved through spatial multiplexing for users with high geometry and through combining for users with low geometry.

6.4.2 Fixed number of antennas per site

In this section, we fix the number of base antennas per site and compare the performance for different numbers of sectors per site. We first study single-

	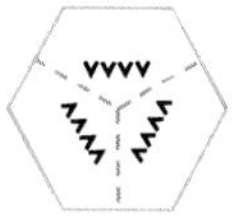	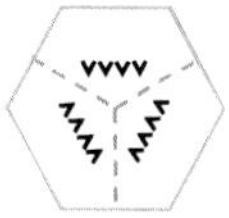
Sectors per site (S)	3	3
Antennas per sector (M)	1, 2, 4	1, 2, 4
Transceiver	SU-MIMO	MU-MIMO
Users per sector (K)	1	$K = M$
Antennas per user (N)	$N = M$	1
Reference SNR (P/σ^2)	30 dB	
Frequency reuse (F)	1	
Transmission mode	Full-power	Full-power, Min. sum power

Fig. 6.18 Parameters for system simulations that show throughput scaling as a function of the number of base antennas per sector M (Section 6.4.1).

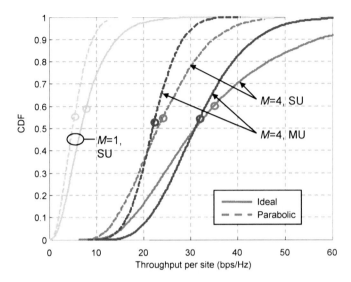

Fig. 6.19 CDF of throughput per site under different sectorization and interference assumptions. Under ideal sectorization, performance improves in going from white to colored interference as a result of Jensen's inequality. Under colored interference, performance degrades in going from ideal to nonideal parabolic sector responses as a result of intersector interference.

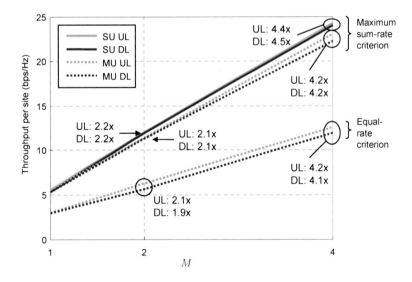

Fig. 6.20 Average throughput (under the proportional-fair criterion) and 10% outage throughput (under the equal-rate criterion) versus M, the number of antennas per sector. Simulation parameters are given by Figure 6.18, under the assumption of parabolic sector responses and colored interference. For both criteria and for both uplink and downlink, throughput scales with M.

user MIMO techniques for six antennas per site. We consider $S = 3$ sectors
per site with $M = 2$ antennas per sector and $S = 6$ sectors per site with
$M = 1$ antenna per sector. We then study multiuser MIMO techniques for
twelve antennas per site with $S = 3, 6, 12$ sectors per site and, respectively,
$M = 4, 2, 1$ antennas per sector.

6.4.2.1 Single-user MIMO, six antennas per site

Fixing the number of antennas and users per site to be six, we consider conventional sectorization ($S = 3$ sectors per site, $M = 2$ antennas per sector,
$K = 2$ users per sector) and higher-order sectorization ($S = 6$ sectors per
site, $M = 1$ antenna per sector, $K = 1$ user per sector). The options and parameters are summarized in Figure 6.21. Sector responses are given by (5.7)
and parameters in Figure 5.5. Multiple antennas at a given base (sector) are
assumed to have the same boresight direction and to be spatially uncorrelated. Therefore they could be implemented using a cross-polarized (Div-1X)
configuration or two widely separated columns (Tx-Div) as discussed in Section 5.5. For the $S = 3$ sector case, we evaluate the performance of Alamouti
transmit diversity, open-loop spatial multiplexing, and closed-loop spatial
multiplexing. We also consider SIMO transmission ($M = 1$) as a baseline
for comparison. Mobile users are assumed to have $N = 2$ antennas, and the
channels are spatially i.i.d. Rayleigh.

Single user per sector
We first consider the distribution of the user rates when there is single user per
sector. The distribution is generated over random user locations, shadowing
realizations, and Rayleigh channel realizations. The CDF for the five cases is
shown in Figure 6.22. Compared to the $S = 3, M = 1$ baseline, the transmit
diversity distribution has a smaller variation so there is a slightly improved
cell-edge performance and lower peak-rate performance. The mean rate is
slightly lower and is a consequence of the Rayleigh channel realizations of
the interfering cells. (If we model the interference using the average power
based on $\alpha_{k,b}$ instead, the mean throughput using transmit diversity would
be slightly higher than the baseline's.) While transmit diversity improves the
link reliability, the reduction in the range of channel variations is actually
detrimental to scheduled systems that rely on large channel variations [157].
Furthermore, in contemporary scheduled cellular networks operating in wide

bandwidth, transmit diversity has limited value because there are other forms of diversity including multiuser diversity, receiver combining diversity and frequency diversity [163].

For CL-MIMO, multiplexing gain is achieved at high geometry and there is a significant improvement in the peak rate compared to the baseline. At low geometries, CSIT allows the transmitter to shift power to the most favorable eigenmode so that CL-MIMO achieves a significant cell-edge rate gain as well. For OL-MIMO, multiplexing gain is achieved at high geometry, and its peak rate nearly matches that of CL-MIMO because their performances are equivalent for asymptotically high geometry. At low geometries, OL-MIMO transmits a single stream using diversity and its performance is equivalent to (2,2) transmit diversity. In practice, because CL-MIMO requires CSIT, the performance advantage over OL-MIMO is reduced if the CSIT becomes less reliable, for example if the mobile speed increases. Using higher order sectorization with $S = 6$ sectors, the geometry distribution is similar to that of the baseline (Figure 6.14), so the rate distribution is similar but with a slightly degraded performance.

	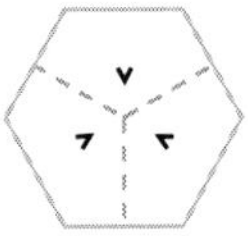	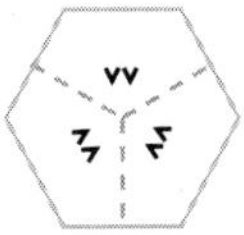	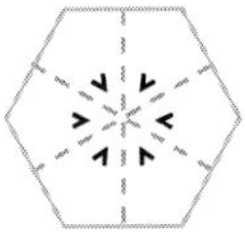
Sectors per site (S)	3	3	6
Antennas per sector (M)	1	2	1
Transceiver	SIMO	Tx Diversity OL-MIMO CL-MIMO	SIMO
Users per sector (K)	2	2	1
Antennas per user (N)	2		
Reference SNR (P/σ^2)	30 dB		
Frequency reuse (F)	1		
Transmission mode	Full-power		

Fig. 6.21 Parameters for system downlink simulations that assume up to 6 antennas per site (Section 6.4.2.1).

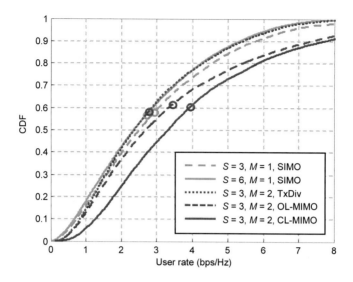

Fig. 6.22 CDF of user rate, 1 user per sector ($N = 2$ antennas per user), for up to 6 antennas per site. CL-MIMO achieves the best performance because it has $M = 2$ antennas with ideal CSIT.

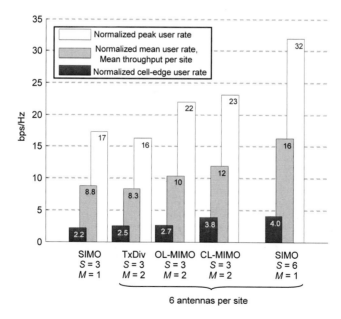

Fig. 6.23 Mean throughput and normalized user rate performance for 6 users per site and up to 6 antennas per site. While the performance *per sector* of higher-order sectorization is inferior to the other options, its performance *per site* is the best.

M users per sector

We emphasize that the rate distributions in Figure 6.22 are from the perspective of the mobile user, and the distributions assume there is a single user per sector. To make fair performance comparisons for different architectures that have different values of S, we consider the performance for a fixed total number of users *per site*. We set the number of users per site to be six and assume an even distribution of users among the sectors. In general, there are M users per sector. For a system with $S = 3$ sectors per site, there are $K = 2$ users per sector, and for system with $S = 6$ sectors per site, there are $K = 1$ users per sector. For the case of $S = 3$, the two users in a sector are served fairly using a round-robin scheduler because their channels are assumed to be static (Section 5.4.1). (If the channels were time varying, then a channel-aware scheduling algorithm could provide multiuser diversity gains over round-robin scheduling.)

For a single user per sector, if the inverse CDF of the user rate R' for a given probability $\alpha \in [0, 1]$ is $F_{R'}^{-1}(\alpha) = x$, then for two users served in a round-robin fashion, the inverse CDF of the user rate R would be $F_R^{-1}(\alpha) = x/2$. Therefore for $K = 2$ users per sector, the CDF of user rate for each of the architectures with $S = 3$ could be obtained by shifting the appropriate curve in Figure 6.22 to the left so that for a given value on the y-axis, the value on the x-axis is reduced by a factor of $1/2$. For example, for SIMO ($S = 3$, $M = 1$) with $K = 2$ users per sector, the cell-edge user rate would be $F_R^{-1}(0.1) = 0.4$ bps/Hz, the peak user rate would be $F_R^{-1}(0.9) = 2.8$ bps/Hz, and the mean user rate would be $\mathbb{E}(R) = 1.5$ bps/Hz.

As defined in Section 5.4.3.1, the normalized peak user rate ($SK \times F_R^{-1}(0.9)$) and normalized cell-edge user rate ($SK \times F_R^{-1}(0.1)$) allow for meaningful comparisons with the mean throughput per site ($SK \times \mathbb{E}(R)$). These statistics are plotted in Figure 6.23. The relative performances of the techniques with $S = 3$ are the same in this figure as they were in 6.22 because the number of users per sector is the same ($K = 2$). However, while the SIMO performance of $S = 6$ is similar to that of $S = 3$ in Figure 6.22, the system-level performance of higher-order sectorization ($S = 6$) is superior with almost a doubling in the values of each metric. The performance of higher-order sectorization is also superior to the CL-MIMO performance for all three metrics. Not only does higher-order sectorization achieve the best performance, its signal processing complexity is less than any of the MIMO techniques because it uses single-antenna transmission and MRC combining.

While single-user spatial multiplexing is known to provide higher peak user rates, it may be surprising that $S = 6$ SIMO achieves the highest normalized peak user rate. The reason is that while the peak rate achieved under $S = 6$ SIMO will be lower for each transmission interval compared to $S = 3$ CL-MIMO, its *normalized* peak rate will be higher because, for a fixed number of users per site, each user under $S = 6$ will be served twice as often.

Another way to understand the gains of sectorization is to note that by doubling the number of sectors per site, the spectral resources are doubled and the performance metrics improve by a factor of two without having to increase the number of antennas per mobile. As shown in Figure 6.24, the mean throughput for SU-SIMO with $S = 6, M = 1, K = 1, N = 2$ (16 bps/Hz) is about twice that of SU-SIMO with $S = 3, M = 1, K = 2, N = 2$ (8.8 bps/Hz). On the other hand, the SU-MIMO system with $S = 3, M = 2, K = 2, N = 2$ uses the same resources as the $S = 6$ SIMO system, but its throughput (12 bps/Hz) is about twice that of the single-user *SISO* system with $S = 3, M = 1, K = 2, N = 1$ (5.3 bps/Hz, from 6.19).

Conclusion: Transmit diversity provides minimal gains in cell-edge rate over single-antenna transmission and is of limited value in contemporary cellular networks which already have other forms of diversity. Closed-loop SU-MIMO is most advantageous with respect to open-loop SU-MIMO at low geometries and at low doppler rates (e.g., pedestrian environments). At high geometries, the performance gap vanishes. Closed-loop (2,2) MIMO provides modest improvements (30%) in peak rate and throughput over (1,2) SIMO. However, for the same number of antennas per site, higher-order sectorization provides even better peak-rate and throughput performance. For systems with many uniformly distributed users, doubling the number of sectors per site doubles the throughput and user metrics.

6.4.2.2 Multiuser MIMO, twelve antennas per site

We now consider 12 antennas per site and the option of more advanced multiuser MIMO techniques. Will higher-order sectorization be competitive with these more advanced MIMO techniques? The transceiver options are summarized in Figure 6.25. Each cell site is split into $S = 3, 6,$ or 12 sectors

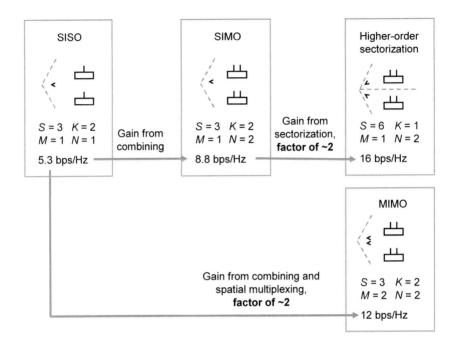

Fig. 6.24 The throughput of higher-order sectorization is approximately double that of the SIMO system, while the throughput of the MIMO system is approximately double that of the SISO system. Therefore for 6 antennas per site, higher-order sectorization results in a higher throughput than the MIMO system.

using, respectively, $M = 4, 2$, and 1 antennas per sector, where multiple antennas for a given sector are assumed to be uncorrelated. Therefore the $M = 2$ and $M = 4$-antenna systems could be implemented using, respectively, a Div-1X and Div-2X antenna configurations (Section 5.5). Both the uplink and downlink performance are studied, and for both cases, the respective capacity-achieving strategy assuming ideal CSIT is used: DPC for the downlink and MMSE-SIC for the uplink.

For $S = 12$, single-antenna reception and transmission are used respectively for the downlink and uplink. As before, the sector responses are given by (5.7) and parameters in Figure 5.5. Twelve users, each equipped with $N = 1$ or 4 antennas, are assigned to each site. For each sector of the $S = 3, 6, 12$-sector configurations, there are respectively $K = 4, 2, 1$ users distributed uniformly in the sector. CL SU-MIMO is used as the baseline for comparison.

	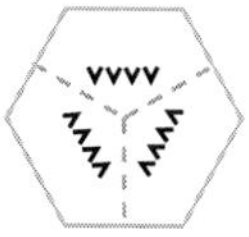		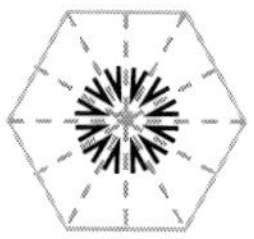
Sectors per site (S)	3	6	12
Antennas per sector (M)	4	2	1
Transceiver	SU-MIMO MU-MIMO	SU-MIMO MU-MIMO	SU-SIMO (DL) SU-MISO (UL)
Users per sector (K)	4	2	1
Antennas per user (N)	1 or 4		
Reference SNR (P/σ^2)	30 dB		
Frequency reuse (F)	1		
Transmission mode	Full-power		

Fig. 6.25 Parameters for uplink and downlink system simulations that assume 12 antennas per site, with $S = 3, 6$ or 12 sectors per site (Section 6.4.2.2). Closed-loop MIMO is used for single-user transmission. Capacity-achieving transceivers are used for MU-MIMO. Single-user SIMO and MISO are used for downlink and uplink transmission, respectively.

Figures 6.26 and 6.27 show the normalized user rate statistics and mean throughput per site for the various transceiver options with $N = 1$ and $N = 4$, respectively. The uplink and downlink results for a given architecture are similar because the underlying geometry distributions are similar when power control is used for the uplink (Figure 5.24). However, as a result of the specific power control parameters used, the uplink has lower cell-edge rates than the downlink. Because of the similarity between the uplink and downlink performance, we focus our discussion on the downlink results.

Single-antenna mobiles, $N = 1$

For now, we consider the downlink results in Figure 6.26 where $N = 1$. We first compare the performance of SU-MISO and MU-MISO for $S = 3, M = 4$. The MU performance is superior to the SU performance because the latter is a special case of the former where transmission is always restricted to a single user at a time. For SU-MISO, the $K = 4$ users per sector are served in

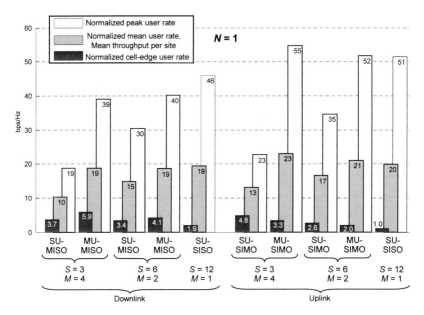

Fig. 6.26 Mean throughput and normalized user rate performance for 12 antennas per site, $N = 1$. For both uplink and downlink, higher-order sectorization with $S = 12$ achieves throughput performance comparable to MU-MIMO with significantly lower signal processing complexity.

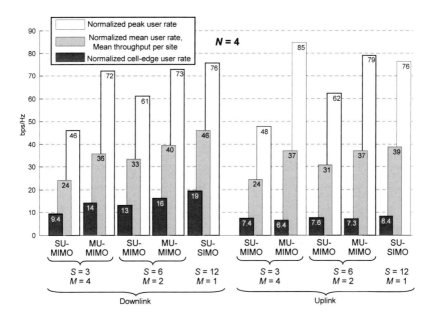

Fig. 6.27 Mean throughput and normalized user rate performance for 12 antennas per site, $N = 4$. Higher-order sectorization performs well, and multiple user antennas increase the cell-edge rate significantly.

a round-robin fashion because the channel is assumed to be static (Section 5.4.1). A single user is served during each transmission interval, but because CSIT is used, the $M = 4$ antennas provide combining gain. For MU-MISO transmission, multiplexing gain can be achieved by serving multiple users simultaneously. Insights for the relative performance of SU- and MU-MISO can be gained from the performance results in Figure 3.15 for $N = 1$. As the SNR increases, the gain of the BC sum-capacity versus the TDMA sum-capacity increases, and this is reflected in Figure 6.26 where the peak rate gains for MU-MISO compared to SU-MISO (39 bps/Hz versus 19 bps/Hz) are higher than the cell-edge rate gains (5.9 bps/Hz versus 3.7 bps/Hz). Overall, the MU transmission results in about a factor of two improvement in mean throughput versus SU transmission (19 bps/Hz versus 10 bps/Hz).

For $S = 6, M = 2$, the MU-MISO performance is superior to the SU-MISO performance because SU transmission is a special case of MU transmission. However, the relative performance advantage is reduced when compared to the $S = 3$ sector case because the multiplexing gain for MU-MISO is reduced from 4 to $\min(M, KN) = \min(2, 4) = 2$.

Considering the performance of higher-order sectorization with $S = 12$ and $M = 1$, we see that its mean throughput (19 bps/Hz) is the same as the throughput for MU-MISO $S = 3$ and MU-MIMO $S = 6$. Higher-order sectorization achieves this performance using a much simpler single-antenna transmission strategy compared to the dirty paper coding strategy required in general for the multiuser transmission. Some insights into the relative performance of MU-MISO $S = 3$ and SU-SISO $S = 12$ can be gained from Figure 3.14 by considering the top BC curve for $N = 1$. This curve shows that the relative performance of the ((4,1),4) BC sum-capacity compared to the (1,4) SU capacity is higher in the low-SNR regime because of the advantage of transmitter combining for the BC. This trend is reflected in the superior cell-edge performance for MU-MISO with $S = 3$ compared to SU-SISO with $S = 12$ (5.9 bps/Hz versus 1.9 bps/Hz). We note that the relative gains in Figure 3.14 should be interpreted as the throughput per sector in the system context. The equivalent mean throughput per site for $S = 12$ and MU-MISO $S = 3$ implies that the mean throughput per sector should be a factor of 4 greater for MU-MISO, and this is consistent with the ratio of average sum rates in Figure 3.14 being between 5 and 3.7 for the relevant range of SNRs.

The concluding observation we make in Figure 6.26 is that the mean throughput and peak rate performance with higher-order sectorization $S = 12$ is comparable to the performance of MU-MISO with fewer sectors per site.

However, the cell-edge performance of higher-order sectorization is inferior.

Multi-antenna mobiles, $N = 4$

If the transmitter has multiple antennas, then multiple antennas at the mobile allow the possibility of spatial multiplexing at high geometry. For $S = 3$, the spatial multiplexing gain for SU-MIMO ($\min(M, N) = 4$) and MU-MIMO ($\min(M, KN) = 4$) are the same, and the gain in peak rate due to multiuser transmission (46 bps/Hz versus 72 bps/Hz) is reduced compared to the MISO case (19 bps/Hz versus 39 bps/Hz). A similar trend can be observed for between SU-MIMO and MU-MIMO for $S = 6$.

For single-antenna transmitters in the case of higher-order sectorization ($S = 12$), multiple antennas at the mobiles can be used to achieve combining gain with respect to the signal from the desired base and interference-oblivious mitigation with respect to interference from other bases. For higher-order sectorization, using $N = 4$ antennas improves the cell-edge user rate by a factor of 10 (19 bps/Hz versus 1.9 bps/Hz). (The same gain was achieved for the cell-edge rate in Figure 6.11 for $F = 1$.)

If the transmitter has multiple antennas, the marginal gains achieved from multiple receive antennas are less. For example, the cell-edge user rate for $S = 3, M = 4$ increases by about a factor of 2.5 (from 3.7 to 9.4 bps/Hz).

Overall, higher-order sectorization achieves the best performance for all three metrics, and the implementation requirements are significantly less than those for MU-MIMO. Higher-order sectorization does not require CSIT and uses simple single-antenna transmission. In contrast, CSIT is required for DPC when $S = 3$ or 6, and the computational complexity of DPC is very high. Furthermore, DPC performance depends on the reliability of the CSIT, and it is much less robust than single-antenna transmission. (For the uplink for $S = 12$, a single antenna is used to detect a single user, and the receiver complexity is lower than for $S = 3$ or 6 where an MMSE-SIC receiver is used.)

Conclusion: If the number of antenna elements per site is fixed at 12 and if the channel angle spread is sufficiently narrow, higher-order sectorization is an effective, low-complexity technique to achieve high average throughput, even when compared to far more sophisticated MU-MIMO techniques. MU-MIMO provides performance gains over SU-MIMO by exploiting multiplexing of streams across multiple users. Multiple antennas at the terminal provide higher gains for users with lower geometry and for bases with fewer antennas. For a given transceiver architecture, the performance between the uplink and downlink is similar because the underlying geometry distributions are similar.

6.4.3 Directional beamforming

The superior performance of higher-order sectorization $S = 12$ described above in Section 6.4.2.2 assumes that the sector response has very good sidelobe characteristics that result in minimal intersector interference. A parabolic sector response (5.7) with parameters in Figure 5.5 can be achieved using a sector antenna implemented using a ULA-4V array (Section 5.5.2). Twelve of these sector antennas are deployed per site, and the result is physically unwieldy and visually unattractive, as shown in Column C of Figure 5.28.

As an alternative, we can create the sector beams electronically using a more compact array, as shown in column G of Figure 5.28, by employing three uniform linear arrays per site. Each array consists of $M = 4$ or 8 antennas with half-wavelength spacing, with each element having the same boresight direction (Figure 5.4, $S = 3$), the same element response (5.7) and the same parameters ($S = 3$ in Figure 5.5. We assume a line-of-sight channel model where base antennas are fully correlated and the relative phase offsets between antennas are a function of the signal direction (2.79). Each array forms four fixed beams with directions shown in Figure 6.28.

Using a Dolph-Chebyshev design criterion (Section 4.2.3), the beamforming weights can be designed to point the top array in Figure 6.28 towards 105. A separate set of weights can be designed for 135 degrees. The beam weights for 45 and 75 degrees can be derived through symmetry, and the weights for these four beams can be used by the other two M-antenna arrays to generate their beams. The resulting beam responses for 105 and 135 degrees are shown in, respectively, the top and bottom subfigures of Figure 6.29. In the top subfigure, the sidelobe characteristics for $M = 4$ and 8 elements are similar, but the beamwidth is narrower with more elements. With $M = 8$, the beamwidth is in fact slightly narrower than the parabolic response beamwidth (5.7) for $S = 12$. The bottom subfigure is for the beams steered towards 135 degrees.

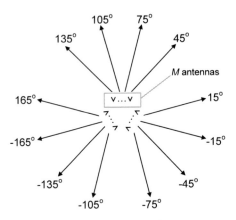

Fig. 6.28 Twelve fixed beams are created per site under directional beamforming. Four beams are formed using a linear array of $M = 4$ or 8 antennas. Directions are measured in degrees with respect to the positive x-axis.

The sidelobe characteristics for $M = 4$ are worse because the desired beam is 45 degrees off the broadside direction of the elements. The beam formed with $M = 8$ elements has better sidelobes and narrower beamwidth.

The twelve beams formed with the ULAs replace the panel sector responses for $S = 12$, and the mean throughput per site versus the channel angle spread is shown in Figure 6.31 for the case of $N = 2$ antennas per user. Compared to the performance of the $M = 4$-element ULA system, the $S = 12$-sector system has about a 25% gain in throughput for zero-degree angle spread as a result of narrower beamwidth and lower sidelobes. However, the ULA performance is similar to that of the $S = 3, M = 4$ system using DPC. Therefore using the same number of antennas per sector, the throughput performance for fully uncorrelated and fully correlated base antennas is similar. Using $M = 8$ elements in the ULA, even though the beams are not the same as the $S = 12$ response, the resulting SINR distributions are similar, and the mean throughput performances are nearly identical for the range of angle spreads considered.

As an alternative to the ULA architecture, the architecture proposed in [164] uses 24 V-pol columns arranged in a circle to form twelve beams for sectorization. As shown in Figure 6.32, each beam is generated using seven adjacent columns, where broadside direction of the center column is pointing in the desired direction of the sector. Due to the symmetry of the circular

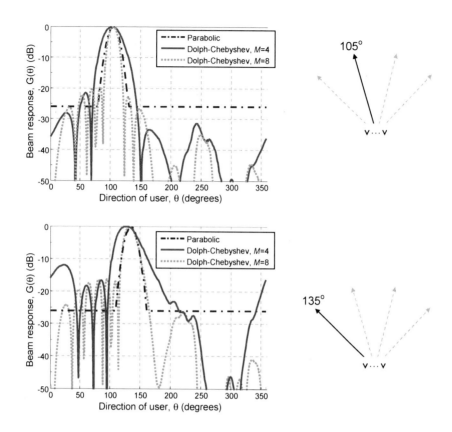

Fig. 6.29 Response of beams formed using uniform linear arrays with $M = 4$ or 8 antennas per sector (with $S = 3$ sectors per site). Four fixed beams are formed per sector in order to approximate the a $S = 12$-sector site using panel antennas to create parabolic sector response. The response of the ULA beam with $M = 4$ in the top subfigure is the same as the Dolph-Chebyshev response in Figure 4.10.

architecture, the response of each beam is identical with respect to the corresponding sector direction, unlike in the case of multiple beams formed with a ULA. It was shown that the measured response of a circular array prototype achieves a similar response as $S = 12$ in Figure 5.5 [164]. Therefore the performance shown in Figures 6.26 and 6.27 for $S = 12$ could be achieved using a much smaller circular array architecture compared to 12 ULA-4V panels.

	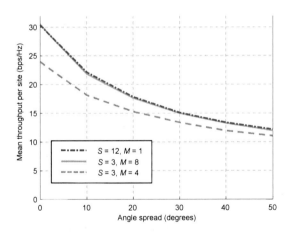	
Sectors per site (S)	3 (12 virtual sectors)	12
Antennas per sector (M)	4, 8	1
Transceiver	SIMO	SIMO
Users per sector	1 per virtual sector	1
Antennas per user (N)	2	
Reference SNR (P/σ^2)	30 dB	
Frequency reuse (F)	1	
Transmission mode	Full-power	

Fig. 6.30 Parameters for system simulations that assume $S = 12$ sectors per site, implemented with either multibeam antenna array or sector antennas (Section 6.4.3). At each site, each of the three multibeam antenna arrays creates four virtual sectors.

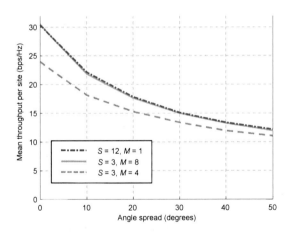

Fig. 6.31 Mean throughput per site versus angle spread. $M = 4$ or 8 antennas are used per sector ($S = 3$) to form beams (see Figure 6.29. Each user has $N = 2$ antennas. The throughput performance with $M = 8$ antennas is identical to the performance with the parabolic $S = 12$ pattern. The throughput performance with $M = 4$ is comparable to the MET ($S = 3$, $M = 4$) performance in Figure 6.26 for zero angle spread.

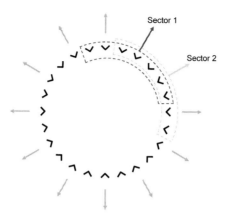

Fig. 6.32 A circular array architecture can be used to create beams for $S = 12$ virtual sectors. The array consists of 24 V-pol columns, and each beam is generated using seven adjacent columns, where broadside direction of the center column is pointing in the desired direction of the sector.

Conclusion: In low angle spread environments, an array of closely spaced base station antennas can be used to create directional beams for sectorization. With enough antennas in the array, the resulting performance is comparable to a system that uses a single antenna for each sector, and because a single array can form multiple beams, the array architecture is more compact. Also, the beams are more flexible and can adapt to non-uniform loading.

6.4.4 Increasing range with MIMO

While we have focused on using MIMO as a means of increasing the spectral efficiency per site for a fixed cell size, it is also possible to tradeoff these gains for increased range. Figure 6.33 shows three user rate distributions to illustrate the range extension benefit in going from a baseline MIMO system to an enhanced MIMO system. The first scenario is the baseline SISO system operating with reference SNR (which we denote as μ) and cell radius d_{ref}. The second scenario is the MIMO system with improved performance with the same reference SNR and cell size. The third scenario is the MIMO system with a reduced reference SNR $\mu' < \mu$ and larger cell size d'_{ref}.

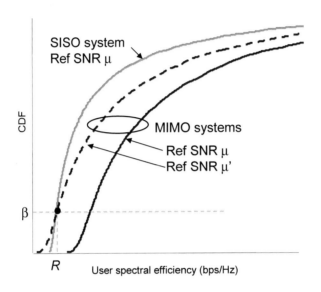

Fig. 6.33 MIMO can be used to extend range in addition to improving spectral efficiency. Using an enhanced System B, the reference SNR can be reduced to μ' to achieve the same cell-edge performance as the baseline System A at reference SNR μ. For a given transmit power, the reduced reference SNR results in a larger cell size.

The new reference SNR μ' is set such that the cell-edge performance of the MIMO system matches that of the SISO system, for example a cell-edge rate R corresponding to an outage probability β. For a given transmit power, a reduced reference SNR corresponds to a network with larger cell radii. Using the definition of the reference SNR in Section 5.2.2, the extended radius is

$$d'_{\text{ref}} = \left(\frac{\mu}{\mu'} \right)^{-\gamma} d_{\text{ref}}, \tag{6.10}$$

where γ is the pathloss exponent. As illustrated in Figure 6.33, the mean user rate (and hence the mean throughput) of the MIMO system with reference SNR μ' will be larger than that of the SISO system. Hence for a given cell-edge performance level, MIMO can achieve a larger mean and higher range compared to SISO. As an alternative to increasing range, one could also use MIMO enhancements to reduce energy consumption by reducing the transmit power for a fixed cell size and for a given $\mu' < \mu$.

Conclusion: One can tradeoff the performance gains in spectral efficiency achieved through multiple antennas for increased range (cell size) or lower transmit power.

6.5 Cellular network, coordinated bases

The results presented so far in this chapter assume that base stations operate independently, and as a result, the performance is interference limited. In this section, we consider the performance of networks with coordination between base stations. As discussed in Section 5.3.2, coordination reduces the intercell interference but requires higher bandwidth and lower latency on the backhaul network to share information between bases. We focus on two coordination techniques: coordinated precoding and network MIMO. The numerical results presented in the section use the equal-rate criterion for fairness (Section 5.4.4).

6.5.1 Coordinated precoding

Under coordinated precoding (CP), the precoding vectors for each base are jointly determined using global knowledge of the user channels. With this knowledge, each base chooses the beamforming weights for its own user(s) while accounting for the interference it causes to users assigned to other bases. We consider a downlink coordinated network using minimum sum-power coordinated precoding, as described in Section 5.4.4.2. Compared to network MIMO precoding, coordinated precoding is attractive because its backhaul bandwidth requirements are far more modest.

Our objective is to compare the spectral efficiency performance between a network with CP and a conventional network with independently operating bases that use conventional (interference-oblivious) precoding. In the conventional network, each base has CSI for only the users it serves. The simulation assumptions are summarized in Figure 6.34. Each site is partitioned into $S = 3$ sectors, and we fix the number of antennas per sector to be $M = 4$, where the sector response is given by (5.7) and Figure 5.5. The number of antennas per user is $N = 1$, and we vary K, the number of users per sector, in the range 1 to 4. We use an equal-rate criterion (but for 10% outage) under which the beamforming weights and user powers are set to minimize the sum of the powers transmitted by all sectors while meeting the prescribed SINR target.

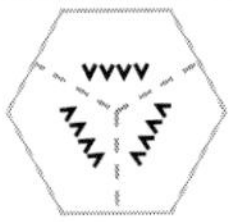

Sectors per site (S)	3	
Antennas per sector	4	
Transceiver	Interf.-oblivious precoding	Coord. precoding
Users per sector (K)	1 to 4	1 to 4
Antennas per user (N)	1	
Reference SNR (P/σ^2)	30 dB	
Frequency reuse (F)	1	
Transmission mode	Minimum sum power	

Fig. 6.34 Parameters for system simulations comparing conventional interference-oblivious precoding with coordinated precoding (Section 6.5.1).

At low reference SNR values, the coordinated and conventional networks have nearly the same spectral efficiency since the network is noise-limited in this regime. However, as we go to higher values of the reference SNR, intercell interference starts to become more significant, and correspondingly, the gains due to coordination increase. In Figure 6.35, the reference SNR is fixed at 30 dB, a value large enough to make the network interference-limited. It shows the spectral efficiency (in bps/Hz per site) for CP and conventional precoding as functions of the number of users per sector.

The spectral efficiency gain due to CP is about 30% with 1 user/sector. Notice that, in this case, each sector uses up only one spatial dimension to serve its user, and has three remaining spatial dimensions to mitigate interference caused to users in other sectors. As the number of users per

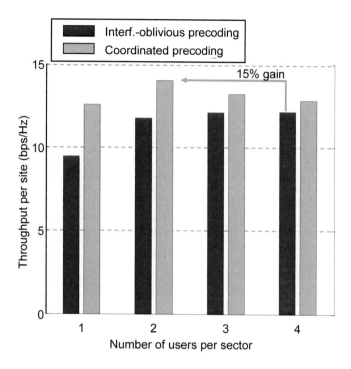

Fig. 6.35 Performance of conventional interference-oblivious precoding versus coordinated precoding. Under CP, each base uses global CSI to optimally balance serving assigned users and mitigating intercell interference. Despite the additional CSI knowledge, it achieves only a 15% improvement in throughput.

sector is increased, the gain in spectral efficiency due to CP becomes smaller, because each sector now has fewer remaining spatial dimensions to mitigate interference. With 4 users/sector, the gain becomes negligibly small since all the available spatial dimensions are being used up for MU-MIMO within the sector.

It is worth noting that the conventional network with $K = 4$ users/sector is quite competitive in the best of the CP network cases (corresponding to 2 users/sector), even though the latter has the advantage of global CSI sharing. This suggests that, in practice, the best complexity-performance tradeoff is attained when each sector simply uses all the available spatial dimensions to serve multiple users without regard to the interference caused to users in other sectors.

Coordinated beam scheduling is a special case of coordinated precoding that shares only scheduling information between bases to avoid beam collisions. It could improve the cell edge performance over conventional interference-oblivious precoding, but its throughput performance can be no better than the coordinated precoding performance and will most likely be only marginally better than conventional precoding.

Conclusion: Using coordinated base stations for downlink transmission, if only CSIT (but not user data) is shared among bases, coordinated precoding with global CSIT provides insignificant gains compared to interference-oblivious precoding under single-base operation. In light of the higher backhaul bandwidth required to share the global CSIT and the higher complexity to compute the joint precoding weights, these gains do not justify the costly upgrade. Coordination techniques that share less information (for example, coordinated scheduling) can perform no better than coordinated precoding with global CSIT.

6.5.2 Network MIMO

With network MIMO coordination, the antennas at multiple base stations act like a single array with spatially distributed elements for the transmission and reception of user signals. As a result, the beamforming weights for each user can be chosen to mitigate the interference from other co-channel users quite effectively. Each user is served by a cluster of coordinated bases, and each base belongs to multiple coordination clusters. Each base has knowledge of the data signals for all users assigned to those clusters and also has global knowledge of the channels between all users and all bases. Coordinated precoding is a special case of network MIMO precoding where the coordination cluster consists of a single base (sector).

Our goal is to understand how the spectral efficiency gain varies with the number of bases being coordinated (cluster size), the reference SNR, and the number of antennas per sector. The simulation assumptions are summarized in Figure 6.36. For the baseline, we assume the site is partitioned into $S = 3$ sectors and each sector has $M = 1, 2$ or 4 antennas, with the sector response given by (5.7) and Figure 5.5. The number of users per sector is M, which, as we discuss later, is the most favorable case for network MIMO. Each user is equipped with a single antenna. For the uplink baseline, each user is detected at a single, optimally assigned base (sector). For the downlink baseline, each user is served by its optimally assigned base using minimum sum-power coordinated precoding.

Under network MIMO, the same antenna architectures as the baseline are used, but antennas are coordinated across each site or across multiple sites. We consider coordination clusters shown in Figure 5.26 of 1, 7, 19, and 61 sites, which we refer to respectively as 0-ring, 1-ring, 2-ring, and 4-ring coordination. Because there are up to twelve antennas per site (in the case of $M = 4$ antennas per sector), coordination occurs for up to 732 antennas in the case of 61 sites.

Figure 6.37 shows the uplink spectral efficiency performance under the equal-rate criterion with different coordination cluster sizes, for $M = 1, 2$ and 4, respectively. The spectral efficiency, measured in bps/Hz per site, is plotted as a function of the reference SNR.

Note that, in each case, the spectral efficiency of the conventional system without any coordination saturates as the reference SNR is increased, indicating that the system is interference-limited. It is this limit that network MIMO attempts to overcome. The following observations can be made:

- The coordination gain increases with the reference SNR P/σ^2 in each case, because interference mitigation becomes more helpful as the level of interference between users goes up relative to receiver noise.
- Intrasite coordination provides modest gains in throughput (less than 20%), but these gains can be achieved with only additional baseband processing at each site compared to the baseline.
- At the low end of the reference SNR range, most of the spectral efficiency gain comes just from 1-ring coordination. This is because most of the interferers that are significant relative to receiver noise are within range of the first ring of surrounding base stations. However, as reference SNR is increased, interferers that are further away start to become significant relative to receiver noise, and therefore it pays to increase the coordination cluster size correspondingly.

The results from the simulations indicate that, in a high-SNR environment, the uplink spectral efficiency can potentially be doubled with 1-ring coordination, and nearly quadrupled with 4-ring coordination. When the user-to-sector-antenna ratio is smaller than 1, the coordination gain will be somewhat lower since, even without coordination, each base station can then use the surplus spatial dimensions to suppress a larger portion of the interference affecting each user it serves. The coordination gain with a user-to-sector-antenna ratio larger than 1 will also be lower, because the composite interference affecting each user at any coordination cluster will then tend towards being spatially white, making linear MMSE beamforming less effective at interference suppression.

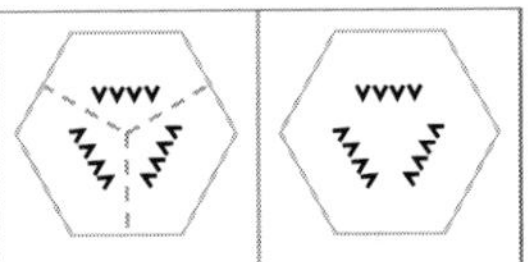

Sectors per site (S)	3	3 (coordinated)
Antennas per sector	1, 2, 4	
Transceiver	Interf.-aware MUD (UL), Coord. precoding (DL)	Network MIMO UL and DL, up to 4 rings coordination
Users per sector (K)	1, 2, 4	
Antennas per user (N)	1	
Reference SNR (P/σ^2)	0 to 30 dB	
Frequency reuse (F)	1	
Transmission mode	Minimum sum power	

Fig. 6.36 Parameters for system simulations comparing network MIMO with conventional operation using independent bases (Section 6.5.2).

Figure 6.38 shows the spectral efficiency achievable on the downlink with different coordination cluster sizes, for $M = 1$, $M = 2$, and $M = 4$, respectively. As on the uplink, the baseline system is interference-limited, and the same observations apply to the downlink. Also, the gains in spectral efficiency from network MIMO (with different cluster sizes) are quite similar on the downlink and uplink.

Conclusion: Using coordinated base stations for downlink transmission, sharing of CSIT and user data allows the possibility of downlink network MIMO, which implements coherent beamforming across multiple sites. Uplink network MIMO implements joint coherent detection

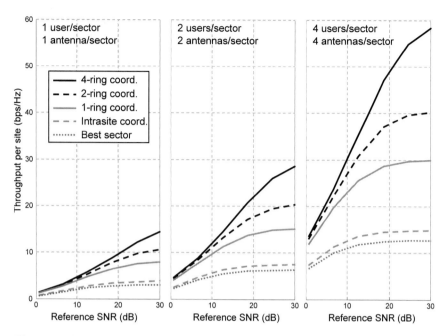

Fig. 6.37 Uplink performance using coordinated multiuser detection across multiple sites (network MIMO). Gains of network MIMO increase as the coordination cluster size increases and as the interference power increases relative to the thermal noise power.

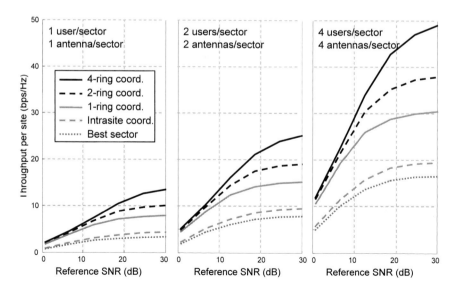

Fig. 6.38 Downlink performance using coordinated precoding across multiple sites (network MIMO). Due to the beamforming duality, the performance trends for the downlink are similar to those for the uplink.

of the users' signals across multiple bases. Network MIMO has the potential to increase both mean throughput and cell-edge rates quite dramatically when compared to the best MIMO baseline with independent bases. As the number of coordinated bases increases, the network MIMO performance improves for a given reference SNR, and the reference SNR value at which the performance becomes interference limited increases. In practice, backhaul limitations and imperfect CSI knowledge due to limited channel coherence reduce the achievable gains.

6.6 Practical considerations

We have made a number assumptions in our cellular system simulations in order to simplify the discussion and to focus on fundamental performance characteristics. In this section, we discuss extensions of the methodology to account for features and impairments found in real-world networks.

Channel estimation

We have assumed that the channel state information is known perfectly at the receiver and, when required, at the transmitter. In Section 4.4, we saw how reference signals are used for channel estimation and how CSI at the base transmitter is obtained through reciprocity in TDD systems or through uplink feedback of PMI and CQI in FDD systems. Acquiring more accurate CSI requires a higher investment in power and bandwidth for the pilot and feedback channels. The overhead also increases when users have higher mobility or when the channel is more frequency selective.

Overhead for orthogonal pilots increases with the number of antennas (for common pilots). This is not the case for dedicated pilots because the number of pilots depends on the number of beams (antenna ports).

We assume that CSI at the receiver (CSIR) can be known ideally by demodulating the reference signals. In practice, these estimates will be noisy, and they can be modeled as additional AWGN [165] [166]. Users with non-ideal CSIR can therefore be modeled as having reduced geometry.

Modulation and coding

The performance metrics used in our numerical results are based on ideal

Shannon capacity bounds. In practice, the data symbols in our system model are digitally modulated symbols mapped from channel encoded information bits. For a given modulation and coding scheme (MCS) indexed by i, the achievable spectral efficiency (in units of bps/Hz) can be written as

$$T_i(\mathsf{SNR}) = (1 - F_i(\mathsf{SNR}))R_i, \tag{6.11}$$

where R_i is the spectral efficiency and $F_i(\mathsf{SNR})$ is the packet error rate (PER) as a function of SNR at the input of the decoder. The PER can be simulated offline and is a function of the encoded block length, the coding rate, the modulation type, and channel characteristics like the mobile speed, and could incorporate practical impairments like channel estimation error. Plotting the achievable spectral efficiency using (6.11) for each MCS, the achievable rate for the overall virtual decoder is given by the outer envelope of the performance curves for all MCSs, as shown in the bottom subfigure of Figure 6.39.

The outer envelope is only a few decibels away from the Shannon bound, and the link performance can be approximated with a 3 dB margin from the bound: $\log_2(1 + \mathsf{SNR}/2)$. This approximation improves as the number of MCSs increases. However, more MCSs require additional CQI feedback bits. Due to constraints on the receiver sensitivity, the modulation is typically limited to 64-QAM so an upper bound on the achievable rate is 6 bps/Hz. Therefore the achievable rate in units of bps/Hz can be approximated as

$$R = \min\left[6, \log_2\left(1 + \frac{\mathsf{SNR}}{2}\right)\right]. \tag{6.12}$$

The link performance of various cellular standards is similar because the encoding is usually based on turbo coding with similar block lengths using 4-QAM, 16-QAM, and 64-QAM.

Wider bandwidths and OFDMA

Our simulations assumed a narrowband channel. In wideband orthogonal frequency-duplexed multiple-access (OFDMA) channels, frequency selective fading could be exploited by scheduling resources in both the time and frequency domains. The precoding and rate could be adapted for each uncorrelated subchannel, but this would require PMI and CQI feedback for each subchannel. A technique known as exponential effective SNR mapping (EESM) can be used to map a set of channel states across a wide bandwidth into a single effective SINR that can be used to predict the achievable rate and

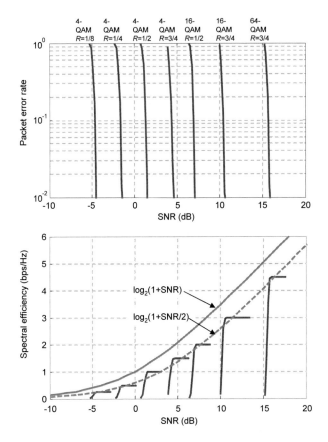

Fig. 6.39 The top subfigure shows the typical packet error rate performance for a stationary user in AWGN. The bottom subfigure shows that the resulting achievable throughput is well-approximated using the Shannon bound with a 3 dB margin. This approximation can be used to model the link performance of the virtual decoder.

block-error performance [167]. In this case only a single CQI is used to characterize the wideband SINR.

Because frequency resources can be allocated dynamically across cells under OFDMA, reduced frequency reuse can be implemented selectively for only users that would benefit most. For example, *fractional frequency reuse* as shown in Figure 6.40 deploys reduced frequency reuse for users with low geometry and universal reuse for all other users. This technique could be implemented through base station coordination or through distributed algorithms for independent bases [159]. However, as was the case for reduced frequency reuse (Section 6.3.3), any gains due to fractional frequency reuse are diminished if multiple antennas are employed at the terminal.

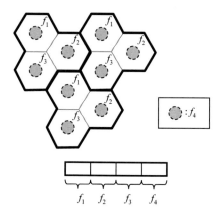

Fig. 6.40 Under fractional frequency reuse, reduced frequency reuse is implemented for users at the cell edges. Users near the bases are served on the same frequency f_4.

Time-varying channels

For high-speed mobiles, the reliability of channel estimation decreases because the time over which the channel is stationary decreases. Also, the channel could change significantly between the time the channel is estimated and the time this information is used for transmission. Techniques that rely on precise channel knowledge such as CSIT precoding are the most sensitive to the mismatch. Codebook precoding is more robust, and CDIT precoding is the most robust because it is based on statistics which vary more slowly than the CSI.

Limited data buffers

Our simulations assume that each user has an infinite buffer of data to transmit. In more realistic simulations, finite buffers and different traffic models could be used to represent classes of services such as voice, streaming audio, streaming video or file transfer.

Synchronization

Coherent transmission and reception requires synchronization so there is no carrier frequency offset (CFO) among the antenna elements of an array. Co-located elements belonging to the same site could be connected to a local oscillator to maintain synchronization. For network MIMO where the elements are distributed spatially, sufficient synchronization could be achieved using commercial GPS (global positioning system) satellite signals for outdoor base stations [161]. Alternatively, without relying on GPS, each base could correct its frequency offset using CFO estimation and feedback from mobiles [168].

Hybrid automatic repeat request (ARQ)

In an ARQ scheme, an error-detecting code such as a cyclic redundancy check (CRC) is used to determine if an encoded packet is received error-free. If so, a positive acknowledgement (ACK) is sent back to the transmitter. If not, a negative acknowledgement (NAK) is sent and another transmission is sent with redundant symbols. The process continues until successful decoding is achieved or until the maximum number of retransmissions occurs. Hybrid ARQ (HARQ) is a combination of forward error-correcting channel coding with an error-detecting ARQ scheme. In current cellular standards, turbo codes or low-density parity-check codes are typically used for channel coding.

When using rate adaptation, there is uncertainty in the SINR knowledge obtained from CQI feedback due to a number of factors. For example, the estimates based on the pilots are inherently noisy, the channel could vary over time, and the interference measured for the CQI feedback could be different from the interference received during the data transmission. The benefit of HARQ is that it allows the scheduler to adapt to this uncertainty by using aggressive rates during the initial transmission followed by retransmissions in response to NAKs. The simulation results in this chapter cannot benefit from HARQ because there is essentially no uncertainty in the channel knowledge. However in practice, HARQ is of great value and is used for almost all data traffic types including latency-sensitive streaming traffic. An examination of HARQ in a MIMO cellular network can be found in [169].

6.7 Cellular network design strategies

One can easily be overwhelmed by the choice of MIMO techniques and design decisions for cellular networks. In this section, we synthesize the insights accrued over the previous chapters and present a unified system design strategy for using multiple antennas in macro-cellular networks. This strategy is meant to provide broad guidelines for deciding among classes of MIMO techniques and to establish a foundation for more detailed simulations that account for practical impairments and cost tradeoffs.

Under the assumption of independent bases (Section 5.3.1), the fundamental techniques for increasing the spectral efficiency per unit area are as follows:

1. **Increase transmit power.** If the transmit powers are set so that the performance is in the noise-limited regime, the spectral efficiency can be improved by increasing the powers until the performance becomes interference limited (Sections 1.3.1 and 6.3.1). In the interference-limited regime, increasing the power does not improve the performance significantly.

2. **Increase cell site density.** By adding cell sites, the throughput per area scales directly with the cell site density, assuming the number of users per cell site remains constant (Section 6.3.2). To improve coverage for indoor environments, indoor bases such as femtocells could be added to offload traffic.

3. **Implement universal frequency reuse.** Given reasonable limits on the transmit power, reusing the spectral resources at each site maximizes the mean throughput per cell with single-antenna transmission (Section 6.3.3). Fractional frequency reuse provides some gains in the cell-edge performance if the mobiles use a single antenna.

4. **Increase spectral efficiency per site.** The spectral efficiency could be increased by improving the link performance, for example with a superior coding and modulation technique or more efficient hybrid ARQ strategy. It could also be increased by using multiple antennas to exploit the spatial dimension of the channel.

These spatial strategies for increasing the spectral efficiency are summarized in Figure 6.41 and described below. In summary, each site should be partitioned into sectors, and MIMO techniques should be implemented within each sector. Additional performance gains could be achieved by coordinating bases and implementing network MIMO.

Sectorization

For both the uplink and downlink, the site should be partitioned into sectors, where the number of sectors is subject to the following three constraints:

1. The angular size of the sector is significantly larger than the angle spread of the channel.
2. The distribution of users is uniform among the sectors.
3. The physical size of the antennas is tolerable.

Within each sector, the following MIMO techniques for the downlink and uplink described below could be used.

Downlink MIMO techniques

Because MIMO techniques for the broadcast channel require knowledge of the CSI at the base station transmitter, the downlink strategies depend on the reliability of this information. If the CSIT is reliable, for example in TDD systems (Section 4.4.1), then CSIT precoding (Section 4.2.1) could be used with either correlated or uncorrelated base antennas. For high-geometry users, optimal multiplexing gain $\min(M, KN)$ can be achieved by using the multiple antennas to transmit multiple streams to one or more users. For low-geometry users, serving a single user at a time using TDMA transmission with a single stream is optimal. Multiple transmit antennas would achieve precoding gain, and multiple receive antennas would achieve combining gain.

If the CSIT is not reliable, then the MIMO strategy depends on the base station antenna correlation. Recall that correlation depends on the spacing of the antennas with respect to the channel angle spread which in turn depends on the relative height of the antennas compared to the surrounding scatterers.

If the base antennas are correlated, CDIT precoding (Section 4.2.2) or CDIT precoding (Section 4.2.3) is effective because a small number of precoding vectors could effectively span the subspace of possible channels. For high-geometry users, multiple users would be served simultaneously to achieve multiplexing gain. Because the transmit antennas are correlated, the channel has rank one, and only a single stream is transmitted to each user. For the case of totally correlated antennas, the multiplexing gain would be no greater than $\min(M, K)$. For low-geometry users, TDMA transmission with a single stream is optimal, and the multiple antennas provide transmitter and receiver combining gains.

If the base antennas are not correlated, then the spatial channel would be too rich for effective codebook/CDIT precoding. If no CSIT is available, then MU-MIMO techniques cannot be used. Using open-loop SU-MIMO, a multiplexing gain of $\min(M, N)$ can be achieved for high geometry (Section 2.3.3). For low-geometry users, TDMA transmission with a single stream is optimal, and multiple receive antennas provide combining gains. Transmit diversity could provide some gains if there are no other forms of diversity such as frequency or multiuser diversity.

The multiplexing gains of SU-MIMO for uncorrelated channels, MU precoding for correlated channels, and MU CSIT-precoding are, respectively, $\min(M, N)$, $\min(M, K)$, and $\min(M, KN)$. For the typical case of $N < M < K$, the performance of these techniques increases roughly in this order, as shown in Figure 6.41. For SU transmission, the rate improves linearly as N increases for high-geometry users due to multiplexing gain (the rate improves linearly as long as $N \le M$) and for low-geometry users due to combining gain (linear gains in post-combining SNR result in linear rate gains at low geometry). For MU transmission, increasing the number of mobile antennas N provides combining gain but does not increase the multiplexing gain for MU transmission if $K > M$. At high geometry, combining gain results in a logarithmic rate increase, but at low geometry, it results in a linear rate increase. Therefore increasing N is beneficial mainly for cell-edge users under MU transmission.

Uplink MIMO techniques

The multiplexing gain of the $((K, N), M)$ MAC is $\min(M, KN)$. For a typical scenario where the number of users exceeds the number of base station antennas $(K < M)$, full multiplexing gain can be achieved by transmitting a single stream from each user, and in this case CSIT is not required by the users. Therefore uplink MIMO performance with full multiplexing gain can be achieved with only CSI at the receiver. The transceiver options for the uplink are simpler than for the downlink because CSI estimates are easier to obtain at the receiver than at the transmitter. If the CSIR is reliable, then multiple users should transmit simultaneously within each sector and capacity-achieving MMSE-SIC could be used. If the CSIR is not reliable, then the more robust MMSE receiver could be used. Multiple terminal antennas could provide precoding gain, but this would require CSIT which could be difficult to obtain in practice.

Coordinated bases

By coordinating the bases to share global CSIT, coordinated downlink precoding could be used to mitigate intercell interference. However, the gains are minimal compared to conventional interference-oblivious transmission with independent bases, and these gains do not justify the cost of acquiring global

CSIT. Additional gains could be achieved using interference alignment, but its performance has not been evaluated in this book.

If the backhaul could be enhanced so that bases share both global CSIT and user data, then downlink coordinated transmission with network MIMO and CSIT precoding could reduce intercell interference to provide more significant gains compared to the baseline. Uplink network MIMO performs joint detection across multiple bases and could provide similar gains. The performance gains of network MIMO increase as the number of coordinated bases increases. However, increasing the coordination cluster size requires additional backhaul resources and acquiring accurate CSI among all transmitter and receiver pairs becomes more difficult.

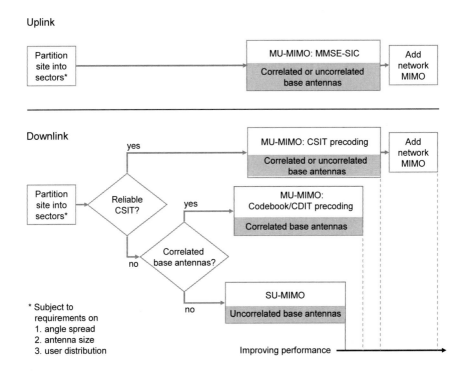

Fig. 6.41 Overview of design recommendations for sectorization and MIMO techniques to increase spectral efficiency per site in a cellular network.

Chapter 7
MIMO in Cellular Standards

Standardization of cellular network technology is required to ensure interoperability between base stations, as well as handsets manufactured by different companies. MIMO standardization at the physical layer focuses on the transmission signaling, describing how information bits are processed for transmission over multiple antennas. Receiver techniques are determined by the terminal or base station vendor and do not need to be standardized.

Since the early 2000s, MIMO techniques have been adopted in cellular standards in parallel with the development of the MIMO theory. The earliest MIMO standardization focused on downlink single-user spatial multiplexing to address the demand for data downloading and higher peak data rates. Recently, there has been more interest in uplink MIMO, multiuser MIMO, and coordinated base techniques. Higher-order sectorization based on fixed sectors does not require standardization because the physical layer is defined for a given sector and is independent of the number of sectors per site.

This chapter gives an overview of conceptual aspects of MIMO techniques for two families of standards. One is based on the Third Generation Partnership Program (3GPP) specification and includes the Universal Mobile Telecommunications Systems (UMTS), Long-Term Evolution (LTE), and LTE-Advanced (LTE-A) standards. The 3GPP specification evolves through a series of *releases*. MIMO techniques were first introduced in UMTS as part of the 3GPP Release 7 specification. LTE is defined in Releases 8 and 9, and LTE-A is defined in Release 10. The other family of cellular standards is based on the IEEE 802.16 specification for broadband wireless access in metropolitan area networks. The IEEE 802.16e standard addresses mobile wireless access in the range of 2 GHz to 6 GHz and is often known as mobile WiMAX (Worldwide Interoperability for Microwave Access, a forum that defines subsets of the 802.16e standard for commercialization). IEEE 802.16m is the evolution of IEEE 802.16e and is sometimes known as Mobile WiMAX Release 2. Both LTE-A and IEEE 802.16m are designed to achieve the International Telecommunication Union (ITU) requirements for fourth-generation cellular standards which include an all-IP packet switched network architecture, peak rates of up to 1 Gbps for low mobility users, and mean throughput per sector of up to 3 bps/Hz [170].

Figure 7.1 summarizes the key features of these standards. Additional details can be found in books [4] [171] [5], tutorial articles [172] [173] [174] [175] [176] and the actual standards documents.

		UMTS	LTE	LTE-A	802.16e	802.16m
Bandwidth		5 MHz	Up to 20 MHz	Up to 100 MHz	Up to 20 MHz	Up to 100 MHz
Multiple access		CDMA	DL: OFDMA UL: SC-FDMA	DL: OFDMA UL: SC-FDMA	OFDMA	OFDMA
DL	SU-MIMO	2 streams	4 streams	8 streams	2 streams	8 streams
	MU-MIMO		4 users	4 users		4 users
UL	SU-MIMO			4 streams		4 streams
	MU-MIMO		4 users	8 users		8 users

Fig. 7.1 Summary of MIMO options for cellular standards. The maximum number of multiplexed streams and users are shown for SU-MIMO and MU-MIMO, respectively.

7.1 UMTS

Single-user spatial multiplexing was first adopted in cellular networks in the UMTS standard for high-speed downlink packet access (HSDPA). The UMTS standard is based on code-division multiple access (CDMA) in 5 MHz bandwidth, and the HSDPA extension achieves high data rates over the high-speed downlink shared channel (HS-DSCH) by transmitting data on up to 15 out of 16 orthogonal spreading codes. The multiple antenna techniques are defined for up to two base station antennas.

HSDPA supports open-loop transmit diversity, known as *space-time transmit diversity* (STTD). A similar technique termed *space-time spreading* (STS) [177] was also adopted as an optional transmit mode for circuit-switched voice in 3GPP2 in 1999 [178]. A block diagram is shown in the top half of Figure 7.2. Information bits are channel- encoded, interleaved, and mapped to QPSK or 16-QAM constellation symbols. The data symbols are multiplexed into multiple substreams (up to 15) which are modulated using mutually orthogonal CDMA spreading codes. These signals are summed and modulated using a cell-specific scrambling sequence. Signals from pairs of consecutive symbol intervals are modulated using Alamouti space-time block coding and transmitted over the two antennas. Because STTD does not depend on the channel realizations, it can be used for terminals with high mobility.

If the terminal is moving slowly, closed-loop transmit diversity and spatial multiplexing can be used to exploit feedback from the terminal. Release 7 of the UMTS standard supports up to two spatially multiplexed streams using up to 64-QAM modulation. While the standard defines the modulation and coding to achieve a peak data rate of 42 Mbps (using two streams with 64-QAM and 0.98 rate coding), this rate is achievable in practice with very low probability, for example, with stationary users located directly in front of the serving base. While aggressive values for peak rate are often defined in standards, they are not necessarily achievable in realistic environments or with significant probability.

Spatial multiplexing is accomplished through precoding, as shown in Figure 7.2. Information bits are first multiplexed into two streams that are separately encoded, interleaved, and mapped to constellation symbols. The two streams have the same modulation and coding, and they are also spread and scrambled using the same sequences. The streams are precoded using a 2×2 matrix $\mathbf{G}$ drawn from a codebook of four unitary matrices:

$$\left\{ \begin{bmatrix} \frac{1}{2} & \frac{1}{2} \\ \frac{1+j}{2\sqrt{2}} & \frac{-(1+j)}{2\sqrt{2}} \end{bmatrix}, \begin{bmatrix} \frac{1}{2} & \frac{1}{2} \\ \frac{1-j}{2\sqrt{2}} & \frac{-(1-j)}{2\sqrt{2}} \end{bmatrix}, \begin{bmatrix} \frac{1}{2} & \frac{1}{2} \\ \frac{-1+j}{2\sqrt{2}} & \frac{-(-1+j)}{2\sqrt{2}} \end{bmatrix}, \begin{bmatrix} \frac{1}{2} & \frac{1}{2} \\ \frac{-1-j}{2\sqrt{2}} & \frac{-(-1-j)}{2\sqrt{2}} \end{bmatrix} \right\}. \tag{7.1}$$

HSDPA spatial multiplexing is sometimes known as dual-stream transmit adaptive array (D-TxAA) transmission and is a generalization of the closed-loop transmit diversity technique known as TxAA. Under TxAA, a single stream is precoded using the weights given by the (re-normalized) left columns of the SM precoding matrices:

$$\left\{ \begin{bmatrix} \frac{1}{\sqrt{2}} \\ \frac{1+j}{2} \end{bmatrix}, \begin{bmatrix} \frac{1}{\sqrt{2}} \\ \frac{1-j}{2} \end{bmatrix}, \begin{bmatrix} \frac{1}{\sqrt{2}} \\ \frac{-1+j}{2} \end{bmatrix}, \begin{bmatrix} \frac{1}{\sqrt{2}} \\ \frac{-1-j}{2} \end{bmatrix} \right\}. \tag{7.2}$$

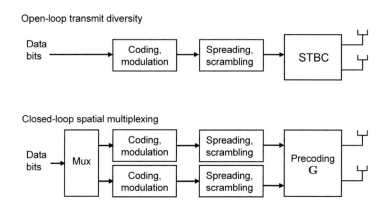

Fig. 7.2 Block diagrams for UMTS MIMO transmission. Up to $M = 2$ base station antennas are supported.

The mobile feeds back precoding matrix indicator (PMI) bits to indicate its preferred weights (Section 5.5.3). Because the codebooks each have four entries, two bits are required for the PMI. We note that SM precoding for two transmit antennas and two streams requires very little PMI feedback, because given one precoding vector, its companion orthogonal vector is immediately known. (A note on terminology: the indicator bits in the UMTS standard are called *precoding control information* (PMI) bits.)

Under spatial multiplexing, the received signal will suffer from inter-code interference as a result of frequency-selective fading and from inter-stream interference as a result of non-ideal unitary precoding. The receiver can account for both effects using multiuser detection techniques combined with a front-end rake receiver or equalizer. A survey of HSDPA receiver designs based on front-end equalization can be found in [179].

7.2 LTE

The Long-Term Evolution (LTE) standard is the next-generation evolution of the UMTS standard based on a downlink orthogonal frequency division multiple access (OFDMA) and an uplink single-carrier frequency-division multiple-access (SC-FDMA) air interface. Compared to UMTS, higher data rates are achieved through wider bandwidths (up to 20 MHz compared to 5 MHz) and higher order spatial multiplexing (up to four streams compared to two). LTE supports up to four antennas at the base. LTE also offers a wider variety of MIMO transmission options including open-loop spatial multiplexing, generalized beamforming, and MU-MIMO.

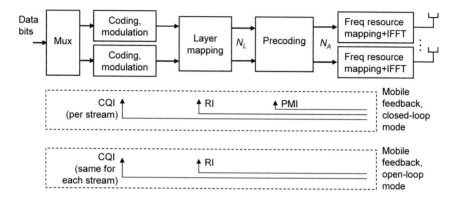

Fig. 7.3 Block diagram for downlink LTE MIMO transmission. Up to $M = 4$ base station antennas are supported. Different precoding is used for closed-loop and open-loop techniques.

On the downlink, three types of single-user transmission are supported: closed-loop transmission (both spatial multiplexing and transmit diversity), open-loop transmission (both spatial multiplexing and transmit diversity) and generalized beamforming. Closed-loop transmission is more suitable for terminals with low mobility and requires more feedback from the terminals than open-loop transmission. LTE also has minimal support for MU-MIMO on both the downlink and uplink.

Downlink SU closed-loop transmission

The transmitter block diagram for LTE downlink spatial multiplexing is shown in Figure 7.3. Information bits are multiplexed into two streams, and each stream is independently coded, interleaved, and mapped to constellation symbols. (In the LTE standard, the term "codeword" is used to denote the independently encoded data streams. To avoid confusion with the precoding codewords, we use the term "stream" instead.) Cell-specific scrambling occurs between the interleaving and modulation to ensure interference randomization between cells. The modulation and coding for each stream can be different and could be based on the CQI feedback from the mobile (Section 5.5.3). If there are four transmit antennas, the two streams are mapped to N_L layers ($N_L = 2, 3, 4$) that correspond to the rank of the spatially multiplexed transmission. The number of layers is suggested by the *rank indicator* (RI) feedback from the mobile. If there are two transmit antennas, the number of layers is $N_L = 2$.

The precoding is implemented with a $N_A \times N_L$ matrix, where N_A is the number of antennas (known as *ports* in the standard). The precoding can be implemented in a frequency-selective manner where different weights are applied across different subbands, or it can be implemented with a single set of weights for the entire transmission band. The precoding is based on the PMI feedback from the mobile. For $N_A = 2$, one of three precoding matrices are used:

$$\mathbf{G} \in \left\{ \begin{bmatrix} \frac{1}{\sqrt{2}} & 0 \\ 0 & \frac{1}{\sqrt{2}} \end{bmatrix}, \begin{bmatrix} \frac{1}{2} & \frac{1}{2} \\ \frac{1}{2} & \frac{-1}{2} \end{bmatrix}, \begin{bmatrix} \frac{1}{2} & \frac{1}{2} \\ \frac{j}{2} & \frac{-j}{2} \end{bmatrix} \right\}. \tag{7.3}$$

The entries for the precoding matrices have been designed so that no complex multiplications are required, only phase rotations by 90, 180, and 270 degrees. For $N_A = 4$, the precoding matrices are chosen from a codebook of size 16. The $4 \times N_L$ matrix corresponds to the N_L columns of Householder matrices given by:

$$\mathbf{G}_n = \mathbf{I}_4 - 2 \frac{\mathbf{u}_n \mathbf{u}_n^H}{\mathbf{u}_n^H \mathbf{u}_n}, \tag{7.4}$$

where n ($n = 0, ..., 15$) is the codebook index, and $\mathbf{u}_n$ is a unique 4×1 generating vector.

After precoding, the data symbols are distributed across the frequency-domain resources to provide frequency diversity. Closed-loop transmit diversity is suitable when the mobile has low mobility and sufficiently high SINR to support multiple streams. If the mobile has low mobility but lower SINR, then closed-loop transmit diversity can be used.

For closed-loop transmit diversity, only a single stream is encoded, and this stream is mapped to a single layer. For $N_A = 2$ antennas, the precoding vector is chosen from a codebook of size 4:

$$\mathbf{g} \in \left\{ \begin{bmatrix} \frac{1}{\sqrt{2}} \\ \frac{1}{\sqrt{2}} \end{bmatrix}, \begin{bmatrix} \frac{1}{\sqrt{2}} \\ \frac{1}{-\sqrt{2}} \end{bmatrix}, \begin{bmatrix} \frac{1}{\sqrt{2}} \\ \frac{j}{\sqrt{2}} \end{bmatrix}, \begin{bmatrix} \frac{1}{\sqrt{2}} \\ \frac{-j}{\sqrt{2}} \end{bmatrix} \right\}. \tag{7.5}$$

Note that HSDPA also used four vectors but with different values. Closed-loop transmit diversity is sometimes known as *codebook-based beamforming*.

For $N_A = 4$ antennas, one of 16 precoding vectors are used, and these vectors are given by the first column of the Householder matrices (7.4).

Downlink SU open-loop transmission

Open-loop spatial multiplexing, sometimes known as large-delay cyclic delay diversity (CDD), is well-suited for high-mobility users whose precoding feedback is not reliable. All layers are constrained to be the same rate, so only a single CQI feedback value is required. After the layer mapping, the layers are precoded to provide diversity across the antennas with the effect of averaging the SINR of each layer. Different precoding matrices are used for frequency different subbands to provide additional frequency diversity.

Open-loop transmit diversity is achieved using Alamouti block coding over the frequency domain. With two transmit antennas, the antenna mapping block-encodes consecutive symbols across consecutive subcarriers, as shown in Figure 7.4. This is known as *space-frequency block coding* (SFBC). With four transmit antennas, the block encoding is similar, but the transmission occurs over alternating antenna pairs to provide additional spatial diversity. This technique is known as SFBC with frequency shift transmit diversity (FSTD).

Given a sequence of encoded symbols $\{b^{(t)}\}$ ($t = 1, 2, \ldots$), each pair of symbols on successive time intervals $b^{(2j-1)}$ and $b^{(2j)}$ ($j = 1, 2, \ldots$) is transmitted over the 2 antennas on intervals $2j - 1$ and $2j$ as follows:

$$\mathbf{s}^{(2j-1)} = \begin{bmatrix} b^{(2j-1)} \\ b^{(2j)} \end{bmatrix} \text{ and } \mathbf{s}^{(2j)} = \begin{bmatrix} -b^{*(2j)} \\ b^{*(2j-1)} \end{bmatrix}.$$

Downlink SU generalized beamforming

If the base station antennas are closely spaced and the channel angle spread is narrow, then directional beamforming can be used to extend the transmission range or to provide transmit combining gain. The beamforming weights in this case would not be drawn from a codebook but would be matched to the direction of the intended user. These non-codebook-based weights could be obtained in an FDD system via estimation of the second order statistics of the uplink channel. Given the beamforming vector $\mathbf{g} \in \mathbb{C}^{M \times 1}$, dedicated reference signals (Section 4.4) are transmitted using $\mathbf{g}$ in order for the terminal to estimate the channel $\mathbf{h}^H \mathbf{g}$. In using the dedicated reference signals, the mobile does not need to have knowledge of $\mathbf{g}$. Generalized beamforming

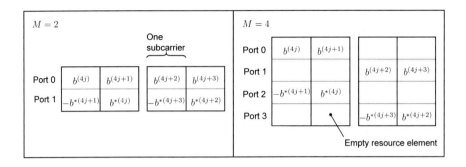

Fig. 7.4 LTE frame structure for open-loop transmit diversity using space-frequency block coding.

supports transmission to only a single user so no multiplexing gains can be achieved.

Downlink MU-MIMO

LTE Release 8 supports beamforming for two users (one layer per user) over the same time-frequency resources. The precoding vectors are drawn from the codebooks for closed-loop transmit diversity. Common reference signals (Section 4.4) are used, and the CQI and PMI feedback sent by each user is computed assuming that it is the only user served. The actual achievable rate would be lower than the predicted rate as a result of interference that occurs during the data transmission phase. Interference is due to interbeam interference from the other user scheduled by the base and intercell interference. The intercell interference could be especially detrimental if the user lies in the direction of an interfering beam, and the potentially significant reduction of SINR is known as the *flashlight effect*.

For LTE Release 9, dedicated reference signals are defined for two layers, enabling non-codebook-based precoding such as zero-forcing for two users. Even though the precoding vectors do not belong to the codebook, the PMI feedback is based on the codebook.

Uplink transmission

On the uplink, SU-MIMO multiplexing is not supported. However, antenna selection is supported for terminals with multiple antennas. In the case of multiuser transmission, a base can schedule more than one user (up to eight) on a given time-frequency resource using SDMA. Users are distinguished from one another using orthogonal reference signals which are transmitted on every fourth time-domain slot.

Two types of reference signal are used for uplink transmission. *Demodulation reference signals* (DRS) are used during data transmission to enable coherent detection. Sounding reference signals (SRS) are transmitted by the terminal for the purpose of resource allocation and estimating the SINR. To enable frequency selective resource allocation, the SRS cover the entire bandwidth during a single transmission interval or by hopping over different subbands over multiple time intervals. In contrast, the DRS are transmitted on only the subbands that carry user data.

7.3 LTE-Advanced

The LTE-Advanced system uses wider bandwidths (up to 100 MHz) and higher multiplexing order for downlink SU-MIMO. In addition, it allows uplink SU-MIMO, and coordinated base station techniques, known as *coordinated multipoint* (CoMP) reception and transmission, are under consideration. At the system level, LTE-A can provide higher spectral efficiency per unit area using heterogenous networks that mix macrocells, femtocells, and relays. LTE-A supports up to eight antennas at the base station with array architectures following the four-antenna configurations in Figure 5.27: eight closely spaced V-pol columns, four columns of closely spaced X-pol columns, and four columns of diversity spaced X-pol columns.

Downlink transmission

On the downlink, higher peak data rates (up to 1 Gbps) can be achieved by leveraging the wider bandwidth and using up to eight layers (with two codewords) for closed-loop SU-MIMO transmission. Of course eight antennas would be required at the base, and at least eight antennas would be required at the terminal. Higher throughput can be achieved using closed-loop MU-MIMO with non-codebook-based precoding for up to four layers and up to two layers per user. For example, the base can serve four users each with one layer or two users each with two layers. LTE-A supports dynamic switching between the SU and MU modes, so the base can decide from frame to frame how to distribute its layers among the users.

Codebooks for two and four transmit antennas are the same as those defined for LTE. For eight transmit antennas, a dual-codebook technique is used such that each codeword is the product of two codewords drawn from two different codebooks. One codebook is designed for the wideband, long-term spatial correlation characteristics. The other codebook is designed for the short-term, frequency-selective properties of the channel. PMI feedback is sent independently for the two codebooks, with the rates scaled accordingly so the latter type occurs more frequently.

Downlink LTE-A supports both common and dedicated reference signals for up to eight transmit antennas. The common reference signals are known as *channel state information reference signals* (CSI-RS). To facilitate CoMP transmission over many bases, up to 40 CSI-RS are defined so that, for example with eight antennas per sector, a frequency reuse of 1/5 could be used to differentiate channels from up to five sectors. The *demodulation reference signals* (DRS) are dedicated reference signals that are modulated using the precoding vector(s). They are used to support non-codebook-based MU-MIMO precoding and can be also be used for codebook-based SU-MIMO precoding. Because there is a DRS associated with each precoded layer, whereas there is a CSI-RS associated with each antenna, channel estimation is more efficient using DRS (for fixed reference signal resources) if the number of layers is less than the number of antennas.

Different types of downlink CoMP transmission strategies are under consideration. Joint processing CoMP is closely related to network MIMO and uses DRS that are jointly precoded across clusters of coordinated bases. Coordinated beam scheduling is also under consideration. In light of the performance results in Chapter 6, minimal gains in throughput compared to conventional single-base precoding would be expected.

Uplink transmission

In addition to supporting uplink MU-MIMO for backwards compatibility with LTE, LTE-A also supports SU-MIMO transmission for two and four antennas. Up to four layers are supported for four antennas. Codebook-based precoding is used for the transmission, deploying the same mechanism of PMI, RI, and CQI feedback from the base to the mobile. However, different codebooks (not based on Householder matrices) are used in order to minimize the peak-to-average power ratio [172]. This is accomplished by allowing each antenna to transmit at most a single layer. The DRS are precoded using the same vectors as the data layers. On the other hand, the SRS are transmitted from each antenna.

CoMP reception on the uplink using network MIMO can be accomplished as a vendor-specific feature and does not require any enhancements in the air-interface standard.

7.4 IEEE 802.16e and IEEE 802.16m

The IEEE 802.16e standard is based on an OFDMA air interface for both the uplink and downlink. Like LTE, it can operate with variable bandwidths from 1.25 to 20 MHz by scaling the FFT size. 802.16e supports both open-loop and closed-loop transmission for two base antennas.

The closed-loop techniques are for single-stream transmission that rely on TDD channel reciprocity for determining precoding weights. Under *maximal ratio transmission*, the precoding weights are computed to maximize the SNR individually on each subcarrier. The channel is assumed to change slowly enough in time so that reliable uplink channel estimates can be made on a per-subcarrier basis. For more rapidly changing channels, a single precoding vector can be applied across the entire band for all subcarriers using *statistical eigenbeamforming* as a form of CDI precoding. In this case, the precoding vector is a function of the largest eigenvector of the channel covariance matrix.

802.16e also supports open-loop transmission over two base antennas. The block diagram is shown in Figure 7.5 for transmit diversity and spatial multiplexing, known respectively as Matrix A and Matrix B transmission. Transmit diversity is achieved using Alamouti space-time block coding. Spatial multiplexing in this case occurs after the coding and modulation, and this option is known as *vertical encoding*. As was the case for open-loop SU-MIMO for LTE, CQI feedback is required for only a single data stream.

IEEE 802.16m is a future enhancement of 802.16e with additional MIMO schemes and with support for up to eight base station antennas and four mobile antennas. Both open and closed-loop SU-MIMO are based on vertical encoding and support up to eight streams on the downlink and up to four streams on the uplink. Downlink MU-MIMO supports up to four users and up to a total of four streams using codebook or CSIT precoding. CDIT precoding can also be implemented using uplink feedback of the downlink channel

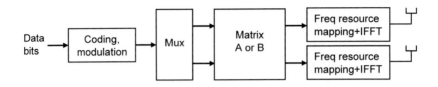

Fig. 7.5 Block diagram for 802.16e open-loop transmission. Matrix A is for Alamouti space-time block coding, and Matrix B is for spatial multiplexing.

covariances to generate a codebook of beamforming vectors that span the likely directions of users. This codebook can be adapted based on additional covariance feedback. If the mobile is slowly moving, then its desired precoding vector is correlated in time. To exploit this correlation, IEEE 802.16m introduces *differential* codebooks so the PMI feedback denotes the incremental change between the previous and current desired codeword.

The IEEE 802.16m standard also supports two types of coordinated beam scheduling. If a base knows the direction of an active user, it can notify an adjacent base of restricted precoding vectors that would cause interference. Conversely, a base can notify an adjacent base of recommended precoding vectors that would not cause interference. These strategies are known respectively as PMI restriction and PMI recommendation. Network MIMO coordinated precoding, for example based on zero-forcing, is also supported.

References

1. G. J. Foschini and M. J. Gans, "On limits of wireless communication in a fading environment when using multiple antennas," *Wireless Personal Communications*, vol. 6, no. 3, pp. 311–335, Mar. 1998.

2. M. Dohler, R. W. Heath, A. Lozano, C. B. Papadias, and R. A. Valenzuela, "Is the PHY layer dead?" *IEEE Communications Magazine*, vol. 49, no. 4, pp. 159–165, Apr. 2011.

3. A. F. Molisch, *Wireless Communications*. John Wiley and Sons, 2011.

4. S. Sesia, I. Toufik, and M. Baker, Eds., *LTE: The UMTS Long Term Evolution*. John Wiley and Sons, 2009.

5. J. G. Andrews, A. Ghosh, and R. Muhamed, *Fundamentals of WiMAX: Understanding Broadband Wireless Networking*. Prentice Hall, 2007.

6. P. Almers, E. Bonek, A. Burr, N. Czink, M. Debbah, V. Degli-Esposti, H. Hofstetter, P. Kyösti, D. Laurenson, G. Matz, A. F. Molisch, C. Oestges, and H. Özcelik, "Survey of channel and radio propagation models for wireless MIMO systems," *EURASIP Journal on Wireless Communications and Networking*, 2007.

7. C.-N. Chuah, J. M. Kahn, and D. Tse, "Capacity of multi-antenna array systems in indoor wireless environment," in *Proceedings of the IEEE Global Telecommunications Conference*, 1998.

8. D. Chizhik, G. J. Foschini, and R. A. Valenzuela, "Capacities of multielement transmit and receive antennas," *Electronics Letters*, vol. 36, no. 13, pp. 1099–1100, 2000.

9. D.-S. Shiu, G. J. Foschini, M. J. Gans, and J. M. Kahn, "Fading correlation and its effect on the capacity of multielement antenna systems," *IEEE Transactions on Communications*, vol. 48, no. 3, pp. 502–513, Mar. 2000.

10. J. P. Kermoal, L. Schumacher, K. Pedersen, P. E. Morgensen, and F. Frederiksen, "A stochastic MIMO radio channel model with experimental validation," *IEEE Journal on Selected Areas in Communications*, vol. 20, no. 6, pp. 1211–1226, Aug. 2002.

11. K. Yu, M. Bengtsson, B. Ottersten, D. McNamara, P. Karlsson, and M. Beach, "Modeling of wide-band MIMO radio channels based on NLoS indoor measurements," *IEEE Transactions on Vehicular Technology*, vol. 53, no. 3, pp. 655–665, May 2004.

12. J. B. Andersen, "Propagation aspects for MIMO channel modeling," in *Space-Time Wireless Systems: From Array Processing to MIMO Communications*, H. Bolcskei,

D. Gesbert, C. B. Papadias, and A.-J. van der Veen, Eds. Cambridge University Press, 2006, pp. 3–22.

13. S. Dossche, S. Blanch, and J. Romeu, "Optimum antenna matching to minimize signal correlation on a two-port antenna diversity system," *Electronics Letters*, vol. 40, no. 19, pp. 1164–1165, 2004.

14. O. N. Alrabadi, C. B. Papadias, A. Kalis, and R. Prasad, "A universal encoding scheme for MIMO transmission using a single active element for PSK modulation schemes," *IEEE Transactions on Wireless Communications*, vol. 8, no. 10, pp. 5133–5143, Oct. 2009.

15. D. Gesbert, H. Bolcskei, D. A. Gore, and A. J. Paulraj, "Outdoor MIMO wireless channels: Models and performance prediction," *IEEE Transactions on Communications*, vol. 50, no. 12, pp. 1926–1934, Dec. 2002.

16. S. Foo, M. Beach, and A. Burr, "Wideband outdoor MIMO channel model derived from directional channel measurements at 2 GHz," in *Proceedings of the International Symposium on Wireless Personal Multimedia Communications*, 2004.

17. M. Debbah and R. Muller, "MIMO channel modeling and the principle of maximum entropy," *IEEE Transactions on Information Theory*, vol. 51, no. 5, pp. 1667–1690, Apr. 2005.

18. A. Sayeed, "Deconstructing multiantenna fading channels," *IEEE Transactions on Signal Processing*, vol. 50, no. 10, pp. 2563–2579, Oct. 2002.

19. C. Balanis, *Advanced Engineering Electromagnetics*. John Wiley and Sons, 1999.

20. C. Cheon, G. Liang, and H. Bertoni, "Simulating radio channel statistics for different building environments," *IEEE Journal on Selected Areas in Communications*, vol. 19, no. 11, pp. 2191–2200, Nov. 2001.

21. J. Ling, D. Chizhik, and R. Valenzuela, "Predicting multi-element receive and transmit array capacity outdoors with ray tracing," in *Proceedings of the IEEE Vehicular Technology Conference*, vol. 1, May 2001, pp. 392–394.

22. W. Lee, "Effects on correlations between two mobile base station antennas," *IEEE Transactions on Communications*, vol. 21, no. 11, pp. 1214–1224, Nov. 1973.

23. J. C. Liberti and T. Rappaport, "A geometrically based model for line-of-sight multipath radio channels," in *Proceedings of the IEEE Vehicular Technology Conference*, April-May 1996, pp. 844–848.

24. O. Norklit and J. Andersen, "Diffuse channel model and experimental results for array antennas in mobile environments," *IEEE Transactions on Antennas and Propagation*, vol. 46, no. 6, pp. 834–843, Jun. 1998.

25. J. Fuhl, A. Molisch, and E. Bonek, "Unified channel model for mobile radio systems with smart antennas," *IEE Proceedings- Radar, Sonar and Navigation*, vol. 145, no. 1, pp. 32–41, Feb. 1998.

26. C. Oestges, V. Erceg, and A. Paulraj, "A physical scattering model for MIMO macrocellular broadband wireless channels," *IEEE Journal on Selected Areas in Communications*, vol. 21, no. 5, pp. 721–729, Jun. 2003.

27. C. Bergljung and P. Karlsson, "Propagation characteristics for indoor broadband radio access networks in the 5 GHz band," in *Proceedings of the International Symposium on Personal, Indoor and Mobile Radio Communications*, 1998.

28. A. F. Molisch, "A generic model for MIMO wireless propagation channels in macro- and microcells," *IEEE Transactions on Signal Processing*, vol. 52, no. 1, pp. 61–71, Jan. 2004.

29. L. M. Correia, *Mobile Broadband Multimedia Networks*. Elsevier, 2006.

30. D. Chizhik, "Slowing the time-fluctuating MIMO channel by beam-forming," *IEEE Transactions on Wireless Communications*, vol. 3, no. 5, pp. 1554–1565, Sep. 2004.

31. J. Ling, U. Tureli, D. Chizhik, and C. B. Papadias, "Rician modeling and prediction for wireless packet data systems," *IEEE Transactions on Wireless Communications*, vol. 7, no. 11, pp. 4692–4699, Nov. 2008.

32. L. M. Correia, *Wireless Flexible Personalised Communications*. John Wiley and Sons, 2001.

33. V. Erceg, L. Schumacher, P. Kyritsi, A. Molisch, D. S. Baum, A. Y. Gorokhov, C. Oestges, Q. Li, K. Yu, N. Tal, B. Dijkstra, A. Jagannatham, C. Lanzl, V. J. Rhodes, J. Medbo, D. Michelson, M. Webster, E. Jacobsen, D. Cheung, C. Prettie, M. Ho, S. Howard, B. Bjerke, L. Jengx, H. Sampath, S. Catreux, S. Valle, A. Poloni, A. Forenza, and T. R. R. W. Heath IEEE P802.11 Wireless LANs, "TGn channel models," IEEE P802.11 Wireless LANs, Technical Report, May 2004.

34. J. Medbo and P. Schramm, "Channel models for HIPERLAN/2," ETSI/BRAN Document No.3ERI085B, 1998.

35. V. Erceg, K. V. S. Hari, M. S. Smith, D. Baum, P. Soma, L. J. Greenstein, D. G. Michelson, S. Ghassemzadeh, A. J. Rustako, R. S. Roman, K. P. Sheikh, C. Tappenden, J. M. Costa, C. Bushue, A. Sarajedini, R. Schwartz, D. Branlund, T. Kaitz, and D. Trinkwon, "Channel models for fixed wireless applications," www.ieee802.org/16/tg3/contrib/802163c-01_29r4.pdf, July 2001.

36. G. Calcev, D. Chizhik, B. Goransson, S. Howard, H. Huang, A. Kogiantis, A. F. Molisch, A. L. Moustakas, D. Reed, and H. Xu, "A wideband spatial channel model for system-wide simulations," *IEEE Transactions on Vehicular Technology*, vol. 56, no. 2, pp. 389–403, Mar. 2007.

37. T. M. Cover and J. A. Thomas, *Elements of Information Theory*. John Wiley and Sons, 1991.

38. E. Telatar, "Capacity of multi-antenna Gaussian channels," *European Transactions on Telecommunications*, vol. 10, no. 6, pp. 585–596, Nov. 1996.

39. D. Tse and P. Viswanath, *Fundamentals of Wireless Communications*. Cambridge University Press, 2005.

40. S. Verdú and S. Shamai, "Spectral efficiency of CDMA with random spreading," *IEEE Transactions on Information Theory*, vol. 45, no. 2, pp. 622–645, Mar. 1999.

41. A. Lapidoth, *A Foundation in Digital Communications*. Cambridge University Press, 2009.

42. R. A. Horn and C. R. Johnson, *Matrix Analysis*. Cambridge University Press, 1994.

43. S. Verdú, *Multiuser Detection*. Cambridge University Press, 1998.

44. X. Wang and H. V. Poor, "Blind multiuser detection: A subspace approach," *IEEE Transactions on Information Theory*, vol. 44, no. 2, pp. 677–690, Aug. 1998.

45. S. T. Chung, A. Lozano, H. C. Huang, A. Sutivong, and J. M. Cioffi, "Approaching the MIMO capacity with a low-rate feedback channel in V-BLAST," *EURASIP Journal on Applied Signal Processing*, pp. 762–771, 2004.

46. M. K. Varanasi and T. Guess, "Optimum decision feedback multiuser equalization with successive decoding achieves the total capacity of the Gaussian multiple-access channel," in *Proceedings of the Asilomar Conference on Signals, Systems and Computers*, 1997.

47. V. Tarokh, N. Seshadri, and A. R. Calderbank, "Space-time codes for high data rate wireless communication: Performance criterion and code construction," *IEEE Transactions on Information Theory*, vol. 44, no. 2, pp. 744–765, Aug. 1998.

48. E. G. Larsson and P. Stoica, *Space-Time Block Coding for Wireless Communications*. Cambridge University Press, 2003.

49. B. Vucetic and J. Yuan, *Space-Time Coding*. John Wiley and Sons, 2003.

50. S. M. Alamouti, "A simple transmitter diversity technique for wireless communications," *IEEE Journal on Selected Areas in Communications*, vol. 16, no. 8, pp. 1451–1458, Oct. 1998.

51. C. B. Papadias, "On the spectral efficiency of space-time spreading schemes for multiple antenna CDMA systems," in *Proceedings of the Asilomar Conference on Signals, Systems and Computers*, 1999, pp. 639–643.

52. S. Sandhu and A. Paulraj, "Space-time codes: A capacity perspective," *IEEE Communications Letters*, vol. 4, pp. 384–386, 2000.

53. A. Lozano, A. M. Tulino, and S. Verdú, "Multiantenna capacity: Myths and reality," in *Space-Time Wireless Systems: From Array Processing to MIMO Communications*, H. Bolcskei, D. Gesbert, C. B. Papadias, and A.-J. van der Veen, Eds. Cambridge University Press, 2006, pp. 87–107.

54. C. Papadias and G. Foschini, "Capacity-approaching space-time codes for systems employing four transmitter antennas," *IEEE Transactions on Information Theory*, vol. 49, no. 3, pp. 726–732, Mar. 2003.

55. C. Yuen, Y. L. Guan, and T. T. Tjhung, *Quasi-Orthogonal Space-Time Block Code*. Imperial College Press, 2007.

56. N. Sharma and C. B. Papadias, "Full rate full diversity linear quasi-orthogonal space-time codes for any transmit antennas," *EURASIP Journal on Applied Signal Processing*, vol. 9, pp. 1246–1256, 2004.

57. B. Hassibi and B. Hochwald, "High-rate codes that are linear in space and time," *IEEE Transactions on Information Theory*, vol. 48, no. 7, pp. 1804–1824, Jul. 2002.

58. M. Sellathurai and S. Haykin, "Turbo-BLAST for wireless communications: Theory and experiments," *IEEE Transactions on Signal Processing*, vol. 50, no. 10, pp. 2538–2546, Oct. 2002.

59. H. El Gamal and A. Hammons, "A new approach to layered space-time coding and signal processing," *IEEE Transactions on Information Theory*, vol. 47, no. 6, pp. 2321–2334, Sep. 2001.

60. M. Trivellato, F. Boccardi, and H. Huang, "On transceiver design and channel quantization for downlink multiuser MIMO systems with limited feedback," *IEEE Journal on Selected Areas in Communications*, vol. 26, no. 8, pp. 1494–1504, Oct. 2008.

61. B. Mondal and R. W. Heath, "Channel adaptive quantization for limited feedback MIMO beamforming systems," *IEEE Transcations on Signal Processing*, vol. 54, no. 12, pp. 4717–4729, Dec. 2006.

62. B. C. Banister and J. R. Zeidler, "Feedback assisted transmission subspace tracking for MIMO systems," *IEEE Journal on Selected Areas in Communications*, vol. 21, no. 3, pp. 452–463, Apr. 2003.

63. ——, "A simple gradient sign algorithm for transmit antenna weight adaptation with feedback," *IEEE Transactions on Signal Processing*, vol. 51, no. 5, pp. 1156–1171, May 2003.

64. J. Yang and D. B. Williams, "Transmission subspace tracking for MIMO systems with low-rate feedback," *IEEE Transcations on Communications*, vol. 55, pp. 1629–1639, 2007.

65. K. K. Mukkavilli, A. Sabharwal, E. Erkip, and B. Aazhang, "On beamforming with finite rate feedback in multiple-antenna systems," *IEEE Transactions on Information Theory*, vol. 49, no. 10, pp. 2562–2579, Oct. 2003.

66. R. Lorenz and S. Boyd, "Robust minimum variance beamforming," in *Robust Adaptive Beamforming*, P. Stoica and J. Li, Eds. John Wiley and Sons, 2006, pp. 1–47.

67. C. R. Murthy and B. D. Rao, "Quantization methods for equal gain transmission with finite rate feedback," *IEEE Transactions on Signal Processing*, vol. 55, no. 1, pp. 233–245, Jan. 2007.

68. V. Raghavan, R. W. Heath, and A. M. Sayeed, "Systematic codebook designs for quantized beamforming in correlated MIMO channels," *IEEE Journal on Selected Areas in Communications*, vol. 25, no. 7, pp. 1298–1310, Sep. 2007.

69. N. Benvenuto and G. Cherubini, *Algorithms for Communications Systems and their Applications*. John Wiley and Sons, 2002.

70. D. J. Love, R. W. Heath, and T. Strohmer, "Grassmannian beamforming for multiple-input multiple-output wireless systems," *IEEE Transactions on Information Theory*, vol. 49, no. 10, pp. 2735–2747, Oct. 2003.

71. B. Mondal and R. W. Heath, "Performance analysis of quantized beamforming MIMO systems," *IEEE Transcations on Signal Processing*, vol. 54, no. 12, pp. 4753–4766, Dec. 2006.

72. A. Ashikhmin and R. Gopalan, "Grassmannian packings for quantization in MIMO broadcast systems," in *Proceedings of the IEEE International Symposium on Information Theory*, 2007.

73. A. F. Molisch, M. Z. Win, Y. Choi, and J. H. Winters, "Capacity of MIMO systems with antenna selection," *IEEE Transactions on Wireless Communications*, vol. 4, no. 4, pp. 1759–1772, July 2005.

74. T. Marzetta, "BLAST training: Estimating channel characteristics for high-capacity space-time wireless," in *Proceedings of the Allerton Conference on Communication, Control and Computing*, Sept. 1999, pp. 958–966.

75. F. Adachi, M. Feeney, J. Parsons, and A. Williamson, "Crosscorrelation between the envelopes of 900 MHz signals received at a mobile radio base station site," *IEE Proceedings- Communications, Radar and Signal Processing*, vol. 133, no. 6, pp. 506–512, Oct. 1986.

76. W. C. Jakes, *Microwave Mobile Communications*. John Wiley and Sons, 1974.

77. A. M. Tulino, A. Lozano, and S. Verdú, "Impact of antenna correlation on the capacity of multiantenna channels," *IEEE Transactions on Information Theory*, vol. 51, no. 7, pp. 2491–2509, Jul. 2005.

78. D. N. C. Tse and S. V. Hanly, "Multiple fading channels. I. Polymatroid structure, optimal resource allocation and throughput capacities," *IEEE Transactions on Information Theory*, vol. 44, no. 7, pp. 2796–2815, Nov. 1998.

79. M. Costa, "Writing on dirty paper," *IEEE Transactions on Information Theory*, vol. 29, no. 3, pp. 439–441, May 1983.

80. H. Weingarten, Y. Steinberg, and S. Shamai, "The capacity region of the Gaussian multiple-input multiple-output broadcast channel," *IEEE Transactions on Information Theory*, vol. 52, no. 9, pp. 3936–3964, Sep. 2006.

81. S. Vishwanath, N. Jindal, and A. Goldsmith, "Duality achievable rates and sum rate capacity of MIMO Gaussian broadcast channels," *IEEE Transactions on Information Theory*, vol. 49, no. 10, pp. 2658–2668, Oct. 2003.

82. W. Yu, W. Rhee, S. Boyd, S. A. Jafar, and J. M. Cioffi, "Iterative water-filling for Gaussian vector multiple-access channels," *IEEE Transactions on Information Theory*, vol. 50, no. 1, pp. 145–152, Jan. 2004.

83. M. Kobayashi and G. Caire, "An iterative waterfilling algorithm for maximum weighted sum-rate of Gaussian MIMO-BC," *IEEE Journal on Selected Areas in Communications*, vol. 24, no. 8, pp. 1640–1646, Aug. 2006.

84. S. Boyd and L. Vandenberghe, *Convex Optimization*. Cambridge University Press, 2004.

85. H. Viswanathan, S. Venkatesan, and H. Huang, "Downlink capacity evaluation of cellular networks with known-interference cancellation," *IEEE Journal on Selected Areas in Communications*, vol. 21, no. 5, pp. 802–811, Jun. 2003.

86. M. Sharif and B. Hassibi, "A comparison of time-sharing, DPC, and beamforming for MIMO broadcast channels with many users," *IEEE Transactions on Communications*, vol. 55, no. 1, pp. 11–15, Jan. 2007.

87. A. M. Tulino, A. Lozano, and S. Verdú, "Multiantenna capacity in the low-power regime," *IEEE Transactions on Information Theory*, vol. 49, no. 10, pp. 2527–2544, Oct. 2003.

88. A. Edelman, "Eigenvalues and condition numbers of random matrices," *SIAM Journal on Matrix Analysis and Applications*, vol. 9, 1988.

89. A. Lapidoth, I. E. Telatar, and R. Urbanke, "On wide-band broadcast channels," *IEEE Transactions on Information Theory*, vol. 49, no. 12, pp. 3250–3260, Dec. 2003.

90. N. Jindal and A. Goldsmith, "Dirty-paper coding vs. TDMA for MIMO broadcast channels," *IEEE Transactions on Information Theory*, vol. 51, no. 5, pp. 1783–1794, May 2005.

91. A. Amraoui, G. Kramer, and S. Shamai, "Coding for the MIMO broadcast channel," in *Proceedings of the IEEE International Symposium on Information Theory*, 2003.

92. H. Viswanathan and S. Venkatesan, "Asymptotics of sum rate for dirty paper coding and beamforming in multiple-antenna broadcast channels," in *Proceedings of the Allerton Conference on Communication, Control and Computing*, 2003, pp. 1064–1073.

93. M. Sharif and B. Hassibi, "On the capacity of MIMO broadcast channels with partial side information," *IEEE Transactions on Information Theory*, vol. 51, no. 2, pp. 506–522, Feb. 2005.

94. T. Y. Al-Naffouri, M. Sharif, and B. Hassibi, "How much does transmit correlation affect the sum-rate scaling of MIMO Gaussian broadcast channels?" *IEEE Transactions on Communications*, vol. 57, no. 2, pp. 562–572, Feb. 2009.

95. J. Hou, J. E. Smee, H. D. Pfister, and S. Tomasin, "Implementing interference cancellation to increase the EV-DO Rev. A reverse link capacity," *IEEE Communications Magazine*, vol. 44, no. 2, pp. 58–64, Feb. 2006.

96. U. Erez and S. ten Brink, "A close-to-capacity dirty paper coding scheme," *IEEE Transactions on Information Theory*, vol. 51, no. 10, pp. 3417–3432, Oct. 2005.

97. U. Erez, S. Shamai, and R. Zamir, "Capacity and lattice strategies for canceling known interference," *IEEE Transactions on Information Theory*, vol. 51, no. 11, pp. 3820–3833, Nov. 2005.

98. A. Khina and U. Erez, "On the robustness of dirty paper coding," *IEEE Transactions on Communications*, vol. 58, no. 5, pp. 1437–1446, May 2010.

99. W. Rhee, W. Yu, and J. M. Cioffi, "The optimality of beamforming in uplink multiuser wireless systems," *IEEE Transactions on Wireless Communications*, vol. 3, no. 1, pp. 86–96, Jan. 2004.

100. P. Wang and L. Ping, "On multi-user gain in MIMO systems with rate constraints," in *Proceedings of the IEEE Global Telecommunications Conference*, 2007.

101. M. L. Honig, Ed., *Advances in Multiuser Detection*. John Wiley and Sons, 2009.

102. B. M. Hochwald, C. B. Peel, and A. L. Swindlehurst, "A vector-perturbation technique for near-capacity multiantenna multiuser communication — Part I: Channel inversion and regularization," *IEEE Transactions on Communications*, vol. 53, no. 1, pp. 195–202, Jan. 2005.

103. ——, "A vector-perturbation technique for near-capacity multiantenna multiuser communication — Part II: Perturbation," *IEEE Transactions on Communications*, vol. 53, no. 3, pp. 537–544, Mar. 2005.

104. F. Boccardi, F. Tosato, and G. Caire, "Precoding schemes for the MIMO-GBC," in *Proceedings of the IEEE International Zurich Seminar on Communication*, 2006.

105. M. Rossi, A. M. Tulino, O. Simeone, and A. M. Haimovich, "Non-convex utility maximization in Gaussian MISO broadcast and interference channels," in *Proceedings of the IEEE International Conference on Acoustics, Speech and Signal Processing*, 2011.

106. G. Dimić and N. D. Sidiropoulos, "On downlink beamforming with greedy user selection: performance analysis and a simple new algorithm," *IEEE Transactions on Communications*, vol. 53, no. 10, pp. 3857–3868, Oct. 2005.

107. T. Yoo and A. Goldsmith, "On the optimality of multiantenna broadcast scheduling using zero-forcing beamforming," *IEEE Journal on Selected Areas in Communications*, vol. 24, no. 3, pp. 528–541, Mar. 2006.

108. F. Boccardi and H. Huang, "Zero-forcing precoding for the MIMO broadcast channel under per-antenna power constraints," in *Proceedings of the IEEE International Workshop on Signal Processing Advances in Wireless Communications*, 2006.

109. Q. H. Spencer, A. L. Swindlehurst, and M. Haardt, "Zero-forcing methods for downlink spatial multiplexing in multiuser MIMO channels," *IEEE Transactions on Signal Processing*, vol. 52, no. 2, pp. 461–471, Feb. 2004.

110. F. Boccardi and H. Huang, "A near-optimum technique using linear precoding for the MIMO broadcast channel," in *Proceedings of the IEEE International Conference on Acoustics, Speech and Signal Processing*, Apr. 2007.

111. A. Bayesteh and A. K. Khandani, "On the user selection for MIMO broadcast channels," *IEEE Transactions on Information Theory*, vol. 54, no. 3, pp. 1124–1138, Mar. 2008.

112. A. M. Tulino, A. Lozano, and S. Verdú, "Capacity-achieving input covariance for single-user multi-antenna channels," *IEEE Transactions on Wireless Communications*, vol. 5, no. 3, pp. 662–673, Mar. 2006.

113. B. Clerckx, G. Kim, and S. Kim, "MU-MIMO with channel statistics-based codebooks in spatially correlated channels," in *Proceedings of the IEEE Global Telecommunications Conference*, 2008, pp. 1–5.

114. D. Hammerwall, M. Bengtsson, and B. Ottersten, "Utilizing the spatial information provided by channel norm feedback in SDMA systems," *IEEE Transactions on Signal Processing*, vol. 56, no. 7, pp. 3278–3295, Jul. 2008.

115. K. Huang, J. G. Andrews, and R. W. Heath, "Performance of orthogonal beamforming for SDMA with limited feedback," *IEEE Transactions on Vehicular Technology*, vol. 58, no. 1, pp. 152–164, Jan. 2009.

116. S. J. Orfanidis, *Electromagnetic Waves and Antennas*. www.ece.rutgers.edu/orfanidi/ewa, Aug. 2010.

117. N. Jindal, "MIMO broadcast channels with finite-rate feedback," *IEEE Transactions on Information Theory*, vol. 52, no. 11, pp. 5045–5060, Nov. 2006.

118. G. Wunder, J. Schreck, P. Jung, H. Huang, and R. Valenzuela, "A new robust transmission technique for the multiuser MIMO downlink," in *Proceedings of the IEEE International Symposium on Information Theory*, 2010.

119. N. Ravindran and N. Jindal, "Limited feedback-based block diagonalization for the MIMO broadcast channel," *IEEE Journal on Selected Areas in Communications*, vol. 26, no. 8, pp. 1473–1482, Oct. 2008.

120. C.-B. Chae, D. Mazzarese, N. Jindal, and R. W. Heath, "Coordinated beamforming with limited feedback in the MIMO broadcast channel," *IEEE Journal on Selected Areas in Communications*, vol. 26, no. 8, pp. 1505–1515, Oct. 2008.

121. P. Viswanath, D. Tse, and R. Laroia, "Opportunistic beamforming using dumb antennas," *IEEE Transactions on Information Theory*, vol. 48, no. 6, pp. 1277–1294, Jun. 2002.

122. D. Avidor, J. Ling, and C. B. Papadias, "Jointly opportunistic beamforming and scheduling (JOBS) for downlink packet access," in *Proceedings of the IEEE International Conference on Communications*, vol. 5, June 2004, pp. 2959–2964.

123. M. Kountouris, D. Gesbert, and T. Salzer, "Enhanced multiuser random beamforming: Dealing with the not so large number of users case," *IEEE Journal on Selected Areas in Communications*, vol. 26, no. 8, pp. 1536–1545, Oct. 2008.

124. F. Rashid-Farrokhi, K. J. R. Liu, and L. Tassiulas, "Transmit beamforming and power control for cellular wireless systems," *IEEE Journal on Selected Areas in Communications*, vol. 16, no. 8, pp. 1437–1450, Oct. 1998.

125. P. Viswanath and D. N. C. Tse, "Sum capacity of the vector Gaussian broadcast channel and uplink-downlink duality," *IEEE Transactions on Information Theory*, vol. 49, no. 8, pp. 1912–1921, Aug. 2003.

126. F. Rashid-Farrokhi, L. Tassiulas, and K. J. R. Liu, "Joint optimal power control and beamforming in wireless networks using antenna arrays," *IEEE Transactions on Communications*, vol. 46, no. 10, pp. 1313–1324, Oct. 1998.

127. G. Caire, N. Jindal, M. Kobayashi, and N. Ravindran, "Multiuser MIMO achievable rates with downlink training and channel state feedback," *IEEE Transactions on Information Theory*, vol. 56, no. 6, pp. 2845–2866, Jun. 2010.

128. D. J. Love, R. W. Heath, V. K. N. Lau, D. Gesbert, B. D. Rao, and M. Andrews, "An overview of limited feedback in wireless communication systems," *IEEE Journal on Selected Areas in Communications*, vol. 26, no. 8, pp. 1341–1365, Oct. 2008.

129. E. Perahia and R. Stacey, *Next Generation Wireless LANs*. Cambridge University Press, 2008.

130. A. D. Wyner, "Shannon-theoretic approach to a Gaussian cellular multiple-access channel," *IEEE Transactions on Information Theory*, vol. 40, no. 6, pp. 1713–1727, Nov. 1994.

131. M. S. Alouini and A. J. Goldsmith, "Area spectral efficiency of cellular mobile radio systems," *IEEE Transactions on Vehicular Technologies*, vol. 48, no. 4, pp. 1047–1066, Jul. 1999.

132. S. Catreux, P. F. Driessen, and L. J. Greenstein, "Data throughputs using multiple-input multiple-output (MIMO) techniques in a noise-limited cellular environment," *IEEE Transactions on Wireless Communications*, vol. 1, no. 2, pp. 226–235, Apr. 2002.

133. N. Marina, O. Tirkkonen, and P. Pasanen, "System level performance of MIMO systems," in *Space-Time Wireless Systems: From Array Processing to MIMO Communications*, H. Bolcskei, D. Gesbert, C. B. Papadias, and A.-J. van der Veen, Eds. Cambridge University Press, 2006, pp. 443–463.

134. W. Choi and J. G. Andrews, "Spatial multiplexing in cellular MIMO-CDMA systems with linear receivers: outage probability and capacity," *IEEE Transactions on Wireless Communications*, vol. 4, no. 4, pp. 2612–2621, Jul. 2007.

135. N. Alliance, "Next generation mobile networks radio access performance evaluation methodology," http://www.ngmn.org, Jan. 2008.

136. O. Tipmongkolsilp, S. Zaghloul, and A. Jukan, "The evolution of cellular backhaul technologies: Current issues and future trends," *Communications Surveys Tutorials, IEEE*, vol. 13, no. 1, pp. 97–113, First quarter 2011.

137. A. Papoulis, *The Fourier Integral and its Applications*. McGraw Hill, 1987.

138. D. Chizhik and J. Ling, "Propagation over clutter: Physical stochastic model," *IEEE Transactions on Antennas and Propagation*, vol. 56, no. 4, pp. 1071–1077, 2008.

139. M. Iridon and D. W. Matula, "Symmetric cellular network embeddings on a torus," in *Proceedings of the International Conference on Computer Communications and Networks*, 1998.

140. V. R. Cadambe and S. A. Jafar, "Interference alignment and the degrees of freedom for the k-user interference channel," *IEEE Transactions on Information Theory*, vol. 54, no. 8, pp. 3425–3441, Aug. 2008.

141. A. S. Motahari, M. A. Maddah-Ali, S. O. Gharan, and A. K. Khandani, "Forming pseudo-MIMO by embedding infinite rational dimensions along a single real line: Removing barriers in achieving the DOFs of single antenna systems," Arxiv preprint, arXiv:0908.2282, 2009.

142. K. Karakayali, G. J. Foschini, and R. A. Valenzuela, "Network coordination for spectrally efficient communications in cellular systems," *IEEE Wireless Communications Magazine*, no. 8, pp. 56–61, Aug. 2006.

143. S. Venkatesan, "Coordinating base stations for greater uplink spectral efficiency in a cellular network," in *Proceedings of the IEEE International Symposium on Personal, Indoor and Mobile Radio Communications*, 2007.

144. K. Balachandran, J. H. Kang, K. Karakayali, and K. M. Rege, "NICE: A network interference cancellation engine for opportunistic uplink cooperation in wireless networks," *IEEE Transactions on Wireless Communications*, vol. 10, no. 2, pp. 540–549, Feb. 2011.

145. C. T. K. Ng and H. Huang, "Linear precoding in cooperative MIMO cellular networks with limited coordination clusters," *IEEE Journal on Selected Areas in Communications*, vol. 28, no. 9, pp. 1446–1454, Dec. 2010.

146. D. Gesbert, S. Hanly, H. Huang, S. Shamai, O. Simone, and W. Yu, "Multi-cell MIMO cooperative networks: A new look at interference," *IEEE Journal on Special Areas in Communications*, vol. 28, no. 9, pp. 1380–1408, Dec. 2010.

147. S. Venkatesan, H. Huang, A. Lozano, and R. Valenzuela, "A WiMAX-based implementation of network MIMO for indoor wireless systems," *EURASIP Journal on Advances in Signal Processing, special issue on Multiuser MIMO transmission with limited feedback, cooperation and coordination*, no. 963547, 2009.

148. S. Venkatesan, "Coordinating base stations for greater uplink spectral efficiency: Proportionally fair user rates," in *Proceedings of the IEEE International Symposium on Personal, Indoor and Mobile Radio Communications*, 2007.

149. A. L. Stolyar, "On the asymptotic optimality of the gradient scheduling algorithm for multiuser throughput allocation," *Operations Research*, vol. 53, no. 1, pp. 12–25, Jan.-Feb. 2005.

150. P. Bender, P. Black, M. Grob, R. Padovani, N. Sindhushayana, and A. Viterbi, "CDMA/HDR: A bandwidth efficient high speed wireless data service for nomadic users," *IEEE Communications Magazine*, vol. 38, no. 7, pp. 70–77, July 2000.

151. A. L. Stolyar, "Greedy primal-dual algorithm for dynamic resource allocation in complex networks," *Queueing Systems*, vol. 54, no. 3, pp. 203–220, 2006.

152. R. Yates and C. Y. Huang, "Integrated power control and base station assignment," *IEEE Transactions on Vehicular Technology*, vol. 44, no. 3, pp. 638–644, Aug. 1995.

153. S. Hanly, "An algorithm for combined cell-site selection and power control to maximize cellular spread spectrum capacity," *IEEE Journal on Selected Areas in Communications*, vol. 13, pp. 1332–1340, Sep. 1995.

154. W. L. Stutzman and G. A. Thiele, *Antenna Theory and Design*. John Wiley and Sons, 1997.

155. S. Shamai and S. Verdú, "The impact of frequency-flat fading on the spectral efficiency of CDMA," *IEEE Transactions on Information Theory*, vol. 47, no. 4, pp. 1302–1327, Aug. 2001.

156. H. C. Huang, H. Viswanathan, A. Blanksby, and M. Haleem, "Multiple antenna enhancements for a high-rate CDMA packet data system," *Journal of VLSI Signal Processing*, vol. 30, pp. 55–69, 2002.

157. H. Viswanathan and S. Venkatesan, "The impact of antenna diversity in packet data systems with scheduling," *IEEE Transactions on Communications*, vol. 52, no. 4, pp. 546–549, Apr. 2004.

158. G. Song and Y. Li, "Utility-based resource allocation and scheduling in OFDM-based wireless broadband networks," *IEEE Communications Magazine*, vol. 43, no. 12, pp. 127–134, Dec. 2005.

159. A. L. Stolyar and H. Viswanathan, "Self-organizing dynamic fractional frequency reuse for best-effort traffic through distributed inter-cell coordination," in *Proceedings of the IEEE International Conference on Computer Communications*, 2009.

160. D. Samardzija and H. Huang, "Determining backhaul bandwidth requirements for network MIMO," in *Proceedings of the European Signal Processing Conference*, 2009.

161. V. Jungnickel, T. Wirth, M. Schellmann, T. Haustein, and W. Zirwas, "Synchronization of cooperative base stations," in *Proceedings of the International Symposium on Wireless Communication Systems*, 2008.

162. B. W. Zarikoff and J. K. Cavers, "Coordinated multi-cell systems: Carrier frequency offset estimation and correction," *IEEE Journal on Selected Areas in Communications*, vol. 28, no. 9, pp. 1490–1501, Dec. 2010.

163. A. Lozano and N. Jindal, "Transmit diversity vs. spatial multiplexing in modern MIMO systems," *IEEE Transactions on Wireless Communications*, vol. 9, no. 1, pp. 186–197, Jan. 2010.

164. H. Huang, O. Alrabadi, J. Daly, D. Samardzija, C. Tran, R. Valenzuela, and S. Walker, "Increasing throughput in cellular networks with higher-order sectorization," in *Proceedings of the Asilomar Conference on Signals, Systems and Computers*, 2010.

165. B. Hassibi and B. M. Hochwald, "How much training is needed in multiple-antenna wireless links?" *IEEE Transactions on Information Theory*, vol. 49, no. 4, pp. 951–963, Apr. 2003.

166. A. Aubry, A. M. Tulino, and S. Venkatesan, "Multiple-access channel capacity region with incomplete channel state information," in *Proceedings of the IEEE International Symposium on Information Theory*, 2010.

167. H. Liu, L. Cai, H. Yang, and D. Li, "EESM based link error prediction for adaptive MIMO-OFDM system," in *Proceedings of the IEEE Vehicular Technology Conference*, 2007, pp. 559–563.

168. B. Zarikoff and J. Cavers, "Multiple frequency offset estimation for the downlink of coordinated MIMO systems," *IEEE Journal on Selected Areas in Communications*, vol. 26, no. 6, pp. 901–912, Aug. 2008.

169. H. Shirani-Mehr, H. Papadopoulos, S. A. Ramprashad, and G. Caire, "Joint scheduling and hybrid-ARQ for MU-MIMO downlink in the presence of inter-cell interference," in *Proceedings of the IEEE International Conference on Communications*, 2010, pp. 1–6.

170. W. Mohr, J. F. Monserrat, A. Osseiran, and M. Werner, "IMT-Advanced and next-generation mobile networks," *IEEE Communications Magazine*, vol. 49, no. 2, pp. 82–83, Feb. 2011.

171. E. Dahlman, S. Parkvall, J. Skold, and P. Beming, *3G Evolution: HSPA and LTE for Mobile Broadband.* Academic Press, 2008.

172. C. S. Park, Y.-P. E. Wang, G. Jongren, and D. Hammarwall, "Evolution of uplink MIMO for LTE-Advanced," *IEEE Communications Magazine*, vol. 49, no. 2, pp. 112–121, Feb. 2011.

173. Q. Li, G. Li, W. Lee, M. Lee, D. Mazzarese, B. Clerckx, and Z. Li, "MIMO techniques in WiMAX and LTE: A feature overview," *IEEE Communications Magazine*, vol. 48, no. 5, pp. 86–92, May 2010.

174. J. Lee, J.-K. Han, and J. Zhang, "MIMO technologies in 3GPP LTE and LTE-Advanced," *EURASIP Journal on Wireless Communications and Networking*, no. 302092, 2009.

175. F. Wang, A. Ghosh, C. Sankaran, P. Fleming, F. Hsieh, and S. Benes, "Mobile WiMAX systems: Performance and evolution," *IEEE Communications Magazine*, no. 10, pp. 41–49, Oct. 2008.

176. K. Balachandran, D. Calin, F.-C. Cheng, N. Joshi, J. H. Kang, A. Kogiantis, K. Rausch, A. Rudrapatna, J. P. Seymour, and J. Sun, "Design and analysis of an IEEE 802.16e-based OFDMA communication system," *Bell Labs Technical Journal*, vol. 11, no. 4, pp. 53–73, 2007.

177. B. Hochwald, T. Marzetta, and C. B. Papadias, "A transmitter diversity scheme for wideband CDMA systems based on space-time spreading," *IEEE Journal on Selected Areas in Communications*, vol. 19, no. 1, pp. 48–60, Jan. 2001.

178. A. Kogiantis, R. Soni, B. Hochwald, and C. Papadias, "Downlink improvement through space-time spreading," IS-2000 standard contribution 3GPP2-C30-19990817-014, Aug. 1999.

179. S. P. Shenoy, I. Ghauri, and D. Slock, "Receiver designs for MIMO HSDPA," in *Proceedings of the IEEE International Communications Conference*, 2008.

Index